The Official Guide to Roger Wilco's Space Adventures

Jill Champion and Richard C. Leinecker

COMPUTE Books
Greensboro, North Carolina

Acknowledgments

Many thanks to Mark Crowe and Scott Murphy (also known as the Two Guys), Anita Greene, Kim Eidson, and the rest of the crew at Sierra for their help and cooperation in putting this book together.

Many thanks to Stephen Levy and Pam Plaut of COMPUTE Books for their inexhaustible (almost) patience, and, of course, for allowing us to document Roger's adventures.

—*The Authors*

Editor: Pam Plaut

Printed in the United States of America
10 9 8 7 6 5 4 3 2 1

Library of Congress Cataloging-in-Publication Data

Champion, Jill
     The official guide to Roger Wilco's space adventures / Jill Champion and Richard C. Leinecker
       p. cm.
     ISBN 0-87455-237-0
     1. Space Quest (Game)   I. Leinecker, Richard C.   II. Title
GV1469.25.S65C47 1991
793.93′2—dc20   91-2845
                                                                                        CIP

COMPUTE Books, 324 West Wendover Ave. Suite 200, Greensboro, North Carolina 27408, is a General Media International company and is not associated with any manufacturer of personal computers.

Contents

Preface

The name's Wilco. Roger Wilco. Okay, so you're no James Bond. You're a . . . well . . . *janitor*. But that's OK—you can still be resourceful. In fact, you'll *have* to be. Although you've never been overly ambitious, you're somehow fated to do more than mop floors and clean toilets. Whatever the reason, you're going to be the final hope of the universe on a number of different occasions—four to be exact—so you might as well accept the fact.

The good part is that you'll be able to count yourself among the greatest heroes of our time—Captain Kirk, Indiana Jones, James Bond, even John Robinson (remember him?). Yes you, Roger Wilco, are about to be more resourceful and cunning than you ever thought possible. If you really put your mind to it, you can be as clever as the rest. There's more than one way to wax a floor, you know.

Have you ever wondered how the other heroes do it? They're always more ingenious than everyone else, surviving the odds time after time. And they somehow pick up on the tiniest of clues—but always, of course, the critical ones—just in the nick of time!

You must have known there's a catch to such adventure prowess. But of course! It's called a *guidebook*—all the big names use them. It's a hint book of sorts—clues you in on what's going to happen so you'll know your options. And it gives you the greatest ideas for using what otherwise seems like a cache of yard-sale junk. With the right guidebook (this one), you'll carry off startling feats and find a way to fly *any* kind of vehicle—even on alien planets. Use your guidebook wisely and nothing will escape your scrutiny.

The other thing you have in your favor is that you're not an extra. They're always the first to go, you know. No sir, you're a regular. Play your cards right and you'll last through *four* episodes of this saga.

You may not be a natural 007, but what difference will that make when you're through with the task at hand? You'll be too much of a hero, right up there in the ranks of Who's Who of Adventure. No one ever has to know your true identity; you'll be known as the world's savviest space quester.

Introduction

As Roger Wilco, you'll walk through each part of the adventures, seeing what he sees and experiencing what he experiences. This is where you'll get the bulk of hints for finding your way through the current setting. Many of the picture captions are clues, so pay close attention to them, too.

You'll discover that there are several ways to handle some situations. Take a dare now and then, but remember to save your game before taking *any* significant course of action. We recommend saving the *same* game at different points under different names. If you've forgotton something or if you make an irreversibly wrong move, you'll be thankful for not having to play the entire game again, even if you do have to replay some of it. Just remember to SAVE YOUR GAME.

Movement Keys: *Space Quests I, II, and III*

If you're using cursor keys rather than a joystick, consider these things:

- Press a cursor key to go; press the same key again to stop.
- Home, PgUp, PgDn, and End move the character diagonally in different directions.
- In *Space Quests I* and *II*, you can set the cursor speed either by typing *slow, normal, fast,* or *fastest* on the command line, or by highlighting the speed in the pull-down menu. In *Space Quest III*, press the equal sign (=) for normal speed, the plus sign (+) for faster speed, or the minus sign (−) for slower speed; or, highlight your selection in the pull-down menu.
- For delicate situations requiring slow movement, go to the slowest speed possible; then rapidly press the cursor key twice to move one step at a time.

Command Syntax: *Space Quests I, II, and III*

One of the best things about *Space Quest* is the large vocabulary and variety of phrases it will accept. For instance, you never need to type a complete sentence. Usually a couple of key words will do—a noun and a verb—and either upper- or lowercase letters will work. Below are some examples. You should experiment yourself to see what *Space Quest* will accept.

- Instead of saying *labion terror beast mating* whistle, just say *whistle*. You may not have *time* to type out a sentence.
- Most objects can be called by a single or shortened name: *bucks* instead of *buckazoids*, *jock* instead of *athletic support-er*, and so on.
- Normally, you would blow a whistle, so don't *eat* it or do anything else out of the ordinary—within reason.
- Sometimes *use* is the correct verb; other times you'll have to be more specific.
- When you *throw* something, you should say where or to whom the object should be thrown.
- The best way to decide which object to use in a certain situation is to look at your inventory and see what's available, or see what you haven't used yet.
- Sometimes you need to *search* rather than just *look at* something to find an object or clue.

Parserless Interface: *Space Quest IV*

Enter *Space Quest IV* and you enter a new realm of complexity and sophistication. Instead of typing in commands, you use a series of icons for walking, looking, handling, talking, smelling, and tasting, and for choosing the items you've collected in your inventory. Volume, speed, and other functions like saving and restarting are controlled by clicking on or choosing the system icon.

Using a mouse to choose icons and to control the game works best, although you can still play *Space Quest IV* with cursor keys or a joystick.

One of the biggest advantages to *Space Quest IV*'s parserless interface is the amount of game control that occurs when you choose a correct action. For instance correctly choosing the hand icon will cause Roger to walk to the object and "take" it.

You're encouraged to use all the icons liberally—try everything. You never know what will happen or what type of response you'll get. Be sure to *look* at everything, too. As always, carefully studying your surroundings can keep you from making fatal errors in judgement! Just remember to save your games often.

Happy space questing!

Chapter 1
Sierra and Its Colorful History

Just south of the sprawling Yosemite National Park in east-central California lies the small city of Oakhurst, home of Sierra On-Line, one of the most successful software publishing companies in the United States. And like few other software companies, Sierra just celebrated its tenth anniversary in 1990. That's quite an achievement, considering the computer software industry itself is barely 11 years old.

What isn't surprising is that this pioneering company is still going strong after a decade of industry ups and downs. Sierra managed to hold on during the mid 1980s when the personal computer industry suffered a near-fatal crash. Scores of businesses—soaring in their successes—were sent plummeting to failure as a result of the fallout. With a little luck and a lot of innovation, Sierra founders Ken and Roberta Williams never gave up the ship. Their ability to bounce back ultimately paid off.

You could say Sierra On-Line was formed on a whim at the Williamses' home in Simi, California, when Roberta merged her flair for playing adventure games with husband Ken's programming talents. The result was *Mystery House*, an Agatha Christie–type murder mystery that incorporated black-and-white drawings of the house with the story. Although there was no animation involved—that was still a few years yet to come—it was the very first computer adventure game to use graphics.

Compared to the sophisticated graphics of today's computer games, the simple line drawings used in *Mystery House* look more like beginning drafts. It's hard to believe that only 11 years ago, those simple drawings spawned a new age in computer adventure text games.

Mystery House was a tremendous success, naturally, because it was the only product of its kind. Of course, in the industry, it was always known that personal computers and software would become more sophisticated as technology grew, but no one had yet put together an adventure game with graphics. That spelled instant success for Ken and Roberta and their newly formed company, On-Line Systems, which later became Sierra On-Line.

After moving their operation (and themselves) to Coarsegold, California, close to Yosemite National Park, the Williamses soon followed *Mystery House* with another adventure game called *The Wizard and the Princess*, also a first of its kind because its graphics were in color. Even though animation was still just a promise for the future at that time, Sierra's games were unique in their use of graphics. As a result, they were very popular with the computer game-playing genre—and they sold like crazy.

With its success, the business eventually outgrew the Coarsegold building. In April 1983, Ken and Roberta moved their company to a larger building in Oakhurst, California, which is still home to Sierra's quality control and public relations departments. Then in January 1984, several months after the move to Oakhurst, Sierra released *King's Quest I*, its first animated, 3-D adventure game.

King's Quest was an extraordinary success, both to the credit of Sierra's unique integration of 3-D graphics with text and music, and to Roberta Williams's original and intriguing story line. After the launch of *King's Quest I*, the Sierra teams created a slew of other original animated adventure games, including the continuing adventures of *King's Quest*, the *Leisure Suit Larry* series, and, of course, the *Space Quest* series. All have had tremendous success in the marketplace.

Sierra On-Line eventually moved to another larger building just outside of Oakhurst, near the Sierra National Forest (there couldn't be a more appropriate backdrop).

With its winning combination of Ken and Roberta Williams and their host of talented programmers, developers, artists, and musicians, Sierra will continue to surprise and please computer game-playing enthusiasts for years to come. In fact, we've probably only just begun to see the best this innovative company has to offer.

Chapter 2
Developing *Space Quests I, II, and III*

Scott Murphy, the programming half of the *Space Quest* design duo, was interested in backpacking in the Sierra mountains when he landed a job with Sierra in the dealer returns department. Mark Crowe, the artistic half, was already working at Sierra in design and marketing when Scott came aboard. Both were interested in the science fiction/space adventure genre and, eventually, after Scott learned to program, these Two Guys from Andromeda, as they're also known, put their heads together, and *Space Quest I* was the result.

The idea for *Space Quest* emerged from the desire to bring about something both humorous and satisfying to science fiction fans. Sierra had so many serious games at the time and needed something with a regular guy—someone who could be a sort of "accidental" hero. Thus, Roger Wilco was born.

Space Quest I followed an exploratory path—since it was one of the first in a new line of Sierra adventure games—so putting the program together took some trial and error. The process turned out to be an entirely evolutionary one with visual arts leading the program's storyline. What the text could say was pretty much limited to what the characters were able to do. To a certain extent, that's still true for Sierra's adventure text games, but the process of creating them is much more structured now.

For *Space Quest I*, the ideas and art were developed as the game was being programmed to see what things *could* happen. Both *Space Quest I* and *II* were programmed in Sierra's own language called AGI, or Adventure Game Interpreter.

Along with the action and art, much of the humor in the original *Space Quest* was created spontaneously while the story was being written. In fact, some of that humor was added after the game was completed and in Beta testing in Sierra's Quality Assurance department (QA). QA is where the program is tested for bugs by walking the character through every possible area and action of the screen, along with typing in all sorts of commands to see the responses. When every conceivable command is anticipated by the testers, strange responses are often returned, some of which are left in the game to add to the humor.

Space Quest II, although still programmed in AGI, followed a more structured course of development than Space Quest I, as did both Space Quests III and IV. Now, all Sierra adventure games are produced following the same type of design document. The story's script is written and then drawn out like a map or flow chart, indicating places, times, events, and actions. These storyboards, similar to those used by advertisers to visualize televison commercials, are produced before the the game is ever programmed, allowing Mark, Scott, and others involved in the game's development to visualize the whole thing before it's ever transferred to the computer screen. The real advantage to following a design document is that so many bugs can be fixed early in the process and at minimal cost.

Space Quest III, also fully scripted in advance of its development cycle, was programmed in SCI, or Sierra's Creative Interpreter. This newer, more sophisticated programming language, made Space Quest III a true 3-D adventure game. Because SCI gave better screen relsolution— twice the resolution of AGI—Mark was able to create more realistic, dimensional graphics, and SCI allowed Mark and Scott to produce better, more creative special effects.

Realistic sound effects created by Sierra's in-house music director, Mark Siebert, were also programmed into Space Quest III—another advantage to Sierra's sophisticated interpreter. Using the MT-32 synthesizer, he was able to manipulate voices into virtually any sound he wanted.

On top of the realistic sound effects, music is essential for creating mood and tone, just as in movies and television. Sierra hired Bob Siebenberg of the rock group Supertramp to expound upon Mark Crowe's original Space Quest musical score. Using a videotape of Space Quest III, Bob viewed the different scenes, keeping in mind Roger's and the other characters' personalities, and wrote music for each scene to set the mood. Having a complete musical soundtrack added another dimension to Space Quest III, setting it apart from any other text adventure game of its time.

Chapter 3
The Making of
Space Quest IV

When we rejoin our favorite protagonist, Roger Wilco—janitor turned space hero—it isn't long before he stumbles his way into the future, courtesy of the Time Rippers. When Roger eventually visits one of his old haunts on the planet Kerona, the difference between past and present is never clearer.

We see a three-dimensional Roger Wilco walking around in the flat, one-dimensional town of Ulence Flats from *Space Quest I*. Ulence Flats contains only 16 colors, contrasting sharply with the 256 colors of *Space Quest IV*. Indeed, the difference between Roger in *Space Quest IV*'s Ulence Flats and *Space Quest I*'s Ulence Flats is so striking, it's almost incredible to believe how much graphic technology has changed. Sierra remains on the leading edge of that technology.

Vectors to Paintings

The earliest versions of *Space Quest* were drawn onscreen by Mark, using Sierra's own graphics tools. Because disk space was so limited, backgrounds were saved as vector images, which contained the instructions for drawing the pictures. When Sierra's art changed to 256 colors along with greater levels of detail, the method changed.

Space Quest IV was the only game in the series to use digitally scanned artwork for the background, although Roger is still drawn onscreen using bitmapped images. While *Space Quest III* was a tremendous visual improvement over the first two games, it still doesn't compare with the realism that makes up the characters and backgrounds of *Space Quest IV*. In fact, *Space Quest IV* has to be seen to be believed.

Let's go back to the beginning to see exactly what creating *Space Quest IV* involved, the best place to start being with the script itself. Before any work is begun, Mark Crowe and Scott Murphy rough out a first draft of the story's plot. Along with this rough draft, or outline, are sketches they make of the different scenes, or *rooms*, that will be involved in acting out the story line.

Rooms are constructed of 16 *priority bands*, much like a horizontal grid, which determine where objects will be placed and where animation will occur. Control lines are added to establish where Roger and other characters can move in the room. For example, priority bands and control lines working together allow Roger to walk through an angular tunnel without passing through its walls. In a scene where Roger blows smoke, these same devices prevent the smoke from passing through walls or going anywhere other than through the passageway where it belongs.

Once a room is constructed, the artwork for that scene—usually an acrylic painting on illustration board—is digitally scanned into the computer's memory. The result is beautiful, lifelike backgrounds of depth and detail, unlike those in any previous *Space Quest* games.

A Character Is Born

As for Roger himself, other than closeups, you normally see a character that was created in a box 33 squares high and about 16 to 18 squares wide. Changing the colors of the squares makes Roger appear to move as his shape and direction are modified. The background color, which is always invisible, assumes whatever color is behind it in the scene, which is how Roger is able to walk around or behind a building rather than through it. Roger and other objects in a scene are animated by displaying a series of drawings in rapid succession, the same way the individual cells of a filmstrip are given "life" when the filmstrip is put into rapid motion. And, just as a motion picture requires hundreds of individual film cells in a strip to achieve smooth movement, Roger's ability to move or perform an activity also requires hundreds of pictures drawn in different formations.

Programmers Take Hold

Once the plot has been approved and agreed upon, the Sierra crew begins the tedious process of programming the story, which involves writing detailed analyses for every room in the game. The easiest way to approach this tremendous task is to divide the game into different sections and concentrate on one section at a time. For instance, in *Space Quest IV*, the Planet Estros would be considered one section.

The plot outline has been drafted, so next on the list is to map out the scene and then decide what will happen in each room of the map. Rooms are numbered and are always referred to by number. For example, a programming note for the room on the planet Estros in which Roger is carried away by the pterodactyl would look something like this:

Estros Buttes Area

ROOM:
315

VIEW(S):
view.303 - vEgoPteraGrab

MUSIC/SFX:
sPteradactylCaw
sEgoYikes

DESCRIPTION OF EVENTS:
The first time Roger enters this room (which will be from room 305) he is picked up immediately by the giant Pterodactyl. NewRoom to 297.

Subsequent visits should he return to the area will be uneventful. Adjoining rooms are 305 to the north and 310 to the west. Things will be controlled by the buttes region regarding the Sequel Police.

Detailed room descriptions such as this one also include where and what music and sound effects should occur. Dialog is included when it occurs, as the following example shows. In this scene, Roger must make his way past the Sequel Police, into the time pod. If the player moves correctly,

Roger makes it to the time pod safely. If not—well, see for yourself.

Super computer Landing Bay Area

ROOM:
530

PIC(S):
530

VIEW(S):
view.530 - vSPHQStuff
view.017 - vSPoliceNoGun

MUSIC/SFX:
sPodMaterialize
sPodDoor

DESCRIPTION OF EVENTS:
This is the Time Pod bay. Roger can arrive here either on foot from 535 or materialize in a Time Pod from 531. He could also come from 531 without materializing if he were to get in and back out without traveling anywhere.

The first time he walks in from 535 he won't appear immediately. Two time pods are parked here. Before Roger enters a third will materialize. Two Sequel Police will exit the pod and walk to the dispatch terminal upscreen. At this point Roger will enter from the right side. Roger will overhear the Sequel Police making their verbal report to the dispatch terminal.

> "I have just completed a scan in the Labion sector of Space Quest II. No sign of presence at this time."

If Roger walks upscreen he will be spotted and terminated by one of the Sequel Police.

Fire upon Roger first delay 2 seconds after end of animation loop.

"HALT"
"Some people just won't follow instructions."
Death message

If he walks downscreen he will be able to enter one of the pods
undetected.

'Hand' on pod - move him over if necessary.
Roger enters pod.
Door closes - sPodDoor.
NewRoom 531.

Although at this point the script is followed as closely as
possible, dialog and actions can still be changed if necessary.
For instance, the description for this room (530) originally
called for two time pods to be parked and a third to materialize. In the final version of the game, only one time pod is
parked, and the second one materializes. The complete set of
progamming notes for one game can be hundreds of pages.

Mission Accomplished

The Sierra staff spends hundreds of hours not only putting
the program together, but also playing the game over and
over to ensure that everything runs smoothly and all bugs
have been cleared.

As in _Space Quest III_, the final touch of the program
comes from _Space Quest IV_'s fully orchestrated original
score. If you've had the forsight to buy a SoundBlaster for
your computer, _Space Quest IV_ sounds like a movie produc-
tion. In fact, the entire effect from the opening credits to the
end of the game is the closest thing you can get to an actual
movie on your computer screen.

Chapter 4
An Interview with Roger

The atmosphere is clear and quiet as we stare back at the dozens of stars—flirting sirens who dazzle us with an occasional wink here, a twinkle there. Deep, wondrous space—truly the final frontier. A reflection in the bay window tells us our host has arrived, and as we turn to greet him and thank him for his generosity in granting us this rare interview, Roger Wilco stumbles over the trash can in the doorway and lands flat on his face.

Red-faced and a little flustered, he picks himself up off the green tiled floor of the employee break room and offers a clumsy apology for his less-than-graceful entrance.

"Gee, I hope you guys won't print that. I was just polishing the floor on level 3—I must have stepped in some wax before it dried," he offers.

They are the very first words to be spoken in an interview with the legendary Roger Wilco, reluctant janitor-turned-hero. We assure him that his secret is safe with us, if he insists, but that we'd rather be able to show the *real* Roger—the human being, the janitor—not some hollow piece of fluff created by half truths and speculation.

Acquiescing, as seems to be his nature, he grants us permission to print whatever we feel is necessary to paint the true picture of his life and personality. We promise not to overuse the "license," and in truth, we're so excited about talking to him in person we have no need to exaggerate—not that we would anyway.

Roger is a man of average height. His figure is less than imposing with an average build. A few lines etch the un-extraordinary—but not unattractive—face, giving him a slightly rugged look that actually belies his young age. But sitting here in front of us, bedecked in rather plain gray coveralls, it's hard to tell whether those lines are from his years of experience or from too much exposure to the more caustic cleaning fluids he uses. In many ways, it's hard to believe this is the same Roger Wilco we've heard so many things about.

11

Authors: Roger, one outcome of your many adventures that has caused, perhaps, the most confusion among friend and foe alike is your decision to return to janitorial work. Shouldn't the name *Roger Wilco* now be synonymous with *carte blanche*?

Roger: *Well . . . (pause) . . . that's a good question—I guess. Um, I'm not sure what you mean by carte blanche. Is that a new word or something?*

Authors: Oh, you know, worldwide acceptance—the red-carpet treatment wherever you go.

Roger: *Well, I do treat a lot of carpets, but all of them aren't red. I've done a lot of blue ones, green ones, brown ones— every color—you name it. And as far as worldwide accep- tance, I think it's safe to say that I've received universal ac- ceptance. I've had very few people or organizations turn down my offers to clean their carpets, no matter what the color.*

Authors: Well, thank you for that very honest and direct an- swer, Roger. Maybe we should rephrase the question some- what. Why have you chosen to stick with your career as a janitor? Wouldn't you fare very well serving as an expert consultant to one of the higher branches of Space Fleet?

Roger: *Oh, I'm sorry . . . (nervous laughter) . . . I guess I didn't quite understand what you were asking. Space Fleet—wow— now that's a thought.*

Uh, actually, I chose to go back to janitorial work because, for one thing, it's what I do best. I've been a janitor for many years, and I'm very good at what I do. I realize a lot of people don't think this is glamorous or anything, and that's true—it isn't glamorous at all. In fact, it's sometimes a very thankless job, but someone has to be dedicated to fighting grime.

Authors: Okaaay, let's move on to another subject, shall we? You originally believed you were rescuing high-level emis- saries when you decided to go to Pestulon and rescue the Two Guys from Andromeda. Were you disappointed?

Roger: *Well, I have to admit that I did expect to be rescuing people who were a little more significant. I mean let's face it—these guys were just software developers who made a lot of money for ScumSoft. Those scumbag software pirates on Pestulon nabbed them for that jerk, Elmo Pug, who forced them to develop big-time software products. Then, rather than risk having his top moneymakers give him the slip, Elmo decided to immobilize the two in jello. Pretty sick setup, huh?*

Authors: Mmm-hmm. Yes, it is.

Roger: *It was quite a surprise to me when I dissolved the jello block holding them captive. I was expecting intelligent government operators. Instead, they just kept asking me all these stupid questions—What's happening? What's going on? Who are you? Where'd you come from? How do we get out of here?—and on and on. There was a time when I thought they'd never shut up.*

Authors: But still, you rescued them. . . .

Roger: *Well, yeah . . . I mean I had to. I'd come all that way to Pestulon—risked my life and everything—I really didn't have a choice. I couldn't just leave them there.*

Authors: So you felt sorry for them?

Roger: *Uh . . . well, I don't know if I'd say I felt SORRY for them—it was more like irritation. Those guys were like gnats swarming around my head, so instead of feeling sympathy, I just wanted to grab my fly swatter. . . . Besides, they would have followed me anyway. Like I said, I HAD to take them.*

Authors: Whatever happened to those Two Guys?

Roger: *Didn't you watch the end of my third adventure?*

Authors: Well, yes, of course. Your adventures are all very well documented. The Two Guys were responsible for adapting your experiences to several computer adventure games; isn't that correct?

Roger: *Yeah, that's right. They went to work for that place we happened to run into—Sierra. As far as I know, they're still there.*

Authors: Obviously, you didn't have any regrets about leaving them there.

Roger: *Oh—gosh, no. Like I said, they were kind of like gnats.*

Authors: Well, you must have gotten together with them at some point to detail your life story; isn't that correct?

Roger: *Well, no. I never actually got back together with them. You see, we sort of had a lot of time to kill after winning that air—space—battle over Pestulon, whatever you want to call it. I just really needed to tell someone what I had been through, and since these guys had just been through an ordeal of their own, who better to talk to? Besides, I certainly had to listen to enough of their chatter.*

Authors: You mean you told them every detail? Every little thing you said—word for word—and they remembered it?

Roger: *Well, yeah . . . (clears throat) . . . um, I just told them everything that happened to me, and I guess they have real good memories. I like to tell things in detail. . . . I guess it was a pretty long story . . . (laughs) . . . no wonder they were so anxious to leave. Of course, I've never agreed with everything they put in those games. Sometimes they make me look pretty darn stupid, you know?*

Authors: But the main thing is that you've come through everything relatively unscathed, and they do show that in the games.

Roger: *That's true; they do show that in the games.*

Authors: Even after your fourth adventure—traveling through time, discovering people you won't even know for years to come—you still managed to survive intact. . . .

Roger: *Uh, I have no knowledge of any time travels.*

Authors: Oh, come on, Roger. You don't remember being captured and nearly tortured by those Estros gals?

Roger: *I have no knowledge of that. In fact, I think those Two Guys from Andromeda made up the whole thing. For darn sure, no bunch of girls captured me. What do I look like, a sissy?*

Authors: That's debatable. After all, you did wear that little black number and the blonde wig. . . .

Roger: *I don't recall that. I have no knowledge of that. I would NEVER wear a dress. Do you hear me? NEVER! That's just another pack of lies fabricated by you-know-who to sell their lousy games. I think they've stretched the license bit just a little too far. . . .*

Authors: All right, all right. Don't get flustered now. We're just asking!

Roger: *Sorry. I guess I did get carried away. No one puts any credence into what those Guys say anyway, right?*

Authors: One last question for you, Roger. You have been called perhaps the most reluctant hero of your time. How do you feel about that?

Roger: *Well, I wouldn't say the most reluctant hero of MY time; I'd say the most reluctant hero of ALL time. Accidental's more like it. I mean I didn't ask for all that stuff to happen to me, so I'd say that statement is pretty accurate. I'm just an average guy.*

Authors: Well, you may be average in some ways, Roger, but let's face it—you were literally hurled into situations that an average person wouldn't be able to handle. You obviously have a gift for extricating yourself from the most undesirable predicaments—some of them life-threatening. Now that isn't exactly an average ability.

Roger: *Well . . . well, I don't know about that. I do think I'm pretty much your average guy—maybe a little bit better-look-*

ing than the average guy . . . (laughs) . . . no, but . . . uh, still, um . . . I don't know. I think somehow when your life is threatened that many times, something else inside of you takes over, and you . . . you're able to do things you normally wouldn't.

You're thrown into circumstances that are so incredible, so unusual, so unlike anything you've been in before, and your mind begins thinking incredibly and unusually, and you become . . . it becomes very resourceful.

Authors: We said only one more question, but please indulge us for just a little longer. Please tell us—and the rest of the world—who was your most formidable opponent?

Roger: *Well, geez . . . uh, gosh. You mean who was the meanest?*

Authors: Well, yes—that's close enough.

Roger: *Gee, I always thought the expression was "forbidden opponent." Guess that shows how smart I am—language arts was never my best subject . . . (laughs) . . . but getting back to the question, they were all mean. I couldn't say that one was meaner than the other—honestly. I couldn't stand any of them—and they couldn't stand me. They were all just a bunch of bad dudes. There's a lot of meanness out in the universe. I wish people would just learn to live together peacefully.*

Authors: Well, thank you for those ponderous words. Just one last question—did you ever become an engineer?

Roger: *Oh, ha ha. Very funny. I am sanitation engineer, you know!*

Authors: Roger, thank you for your time and cooperation in talking to us here today. Good luck, and stay out of trouble.

Roger: *You're welcome. Thank YOU.*

We feel a certain exhilaration in the air as our reluctant hero leaves us, for we know these past moments have been history in the making. He may be "just average" in looks and intellect, and gawky and gullible in character, but his capacity for conviction extends well above the norm.

His fortitude and resourcefulness wax equally bright—no pun intended—and by their very nature must be expressed. That's why we believe this unpretentious, self-effacing champion won't be washing windows forever. Roger Wilco must live the life he was born to live—that of a hero. And that's no accident . . . CRASH CLANG BANG* . . . but that was!

*Rushing to the source of the clamor, we find Roger Wilco sitting on the floor in the hallway. Around him, on top of him, and underneath him are several sandwich remnants; glops of mustard, mayonnaise, and catsup; a puddle of coffee; and various other items from the sandwich cart he nearly leveled. In his haste to return to level 3, he tells us, he forgot about the wax he stepped in earlier—and the rest, as we say, is history.

Chapter 5

The Sarien Encounter:
Where No Janitor Has
Gone Before

Forget your mop and broom, Roger Wilco, you're about to become the world's last hope for survival. Lucky for you the janitorial closet made such a nice snoozing post; otherwise, you'd be wiped out like the rest of the crew on the spaceship *Arcada*. That's right—the *rest* of the crew. Once a certain scientist dies (and he will very soon), you'll be the only human being left alive on the spaceship. It will then be up to you to find your way out, pick up a few things along the way, and get the heck off the *Arcada*.

It might help you to know that the Sariens have stolen a certain piece of sensitive equipment—known as the Star Generator—from the *Arcada*. Your only purpose in life now (what little of it you have left) is to find and destroy the Generator before those malevolent little creeps figure out how to use it.

Just remember, while you're on the *Arcada*, you aren't the only *thing* walking around the ship. If you're smart, you'll do a disappearing act the moment you hear footsteps. Sariens shoot to kill.

Total Possible Game Points: **202**

The Spaceship *Arcada*

Mission: Find your way to the escape pod and leave the *Arcada*.
Total Points: 36

The Spaceship Arcada

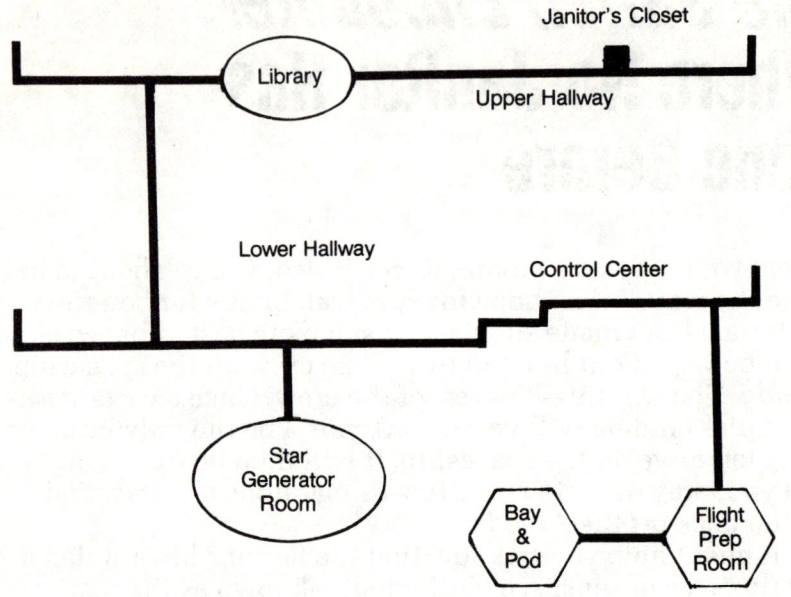

Inventory

instructional cartridge
keycard
spacesuit
translator device

Arcada Upper Level

Hey, what the . . . what *is* that noise? Oh, no. Don't tell me the stupid intruder alarm went off again. I told them not to order that junk from Ronko, but no one ever listens to me. I'm just the *janitor*. Of course, I'm always the one who has to fix everything, too. Darn it. And my dream was just getting good. OK, closet, don't go away—I'll be back later. For now it's out into the hall and. . . . Gee, that's strange. Where is everybody?

I think I'll just take a little hike down the hall to the data library and see if anyone's in there. Now that I think about it, I sure hope that ol' Ronko gadget is just giving off another false alarm. I'd rather fix it again than come face to face with some crazy intruders. I mean, what would I do, mop them to death?

OK, just keep your cool, Wilco. You know there's going to be some nerd in here studying up on. . . . Oh, great. It's empty. No nerds in here. No *anybody* in here. This is starting to get spooky. Too bad this tin can doesn't have an intercom system. Here we are, carrying the Star Generator, the most valuable piece of equipment in the entire universe, and the host ship doesn't even have an intercom. They probably. . . . *Oh, my gosh!* The STAR GENERATOR!! What if someone's taken the Star Generator?! I'd better get over to the Star Generator room fast.

Uh-oh. Are those footsteps I hear? They don't sound like *regular* footsteps. Well, I'm not waiting around to find out who they belong to. I'm getting the heck out of the hall and back into the library.

Did I glimpse a red uniform? Who the heck would wear a red uniform? Oh, no. It can't be. Of course—Sariens. Only Sariens would wear such a tacky getup. They're here for the Star Generator. What am I going to do?

Someone's coming. A Sari— "Professor! Oh, thank goodness. Professor, sir, boy am I glad to see you! I just came through that door. Why didn't I see you in the hall?

"Oh, never mind. I'm Roger Wilco, the janitor. What's going on? I thought I was the only one left alive on this. . . . Sir, are you all right?" Of course, he isn't all right, blip-head—he's grasping his heart and falling to the ground!

"What did you say? Astral what? Mister . . . uh, Doctor-. . . Professor, sir, please don't die. Please don't leave me! I didn't understand what you said. What is an *astral thingama-jig?!* You can't die without telling me. I'm just a janitor! Do you hear me? A JANITOR! I don't know anything about space stuff. I only know about Endust and Lysol!"

Oh, no. He's a goner. Now *I'm* a goner for sure. At least I had the good sense to come back to the library. Otherwise, I might never have seen the professor.

Now to figure out what he's talking about—and fast. Since we're in a cartridge library, I wonder if . . . yeah, that's it. Maybe he was talking about a cartridge! But there are thousands of cartridges in here. Let's see—*It Came from Outer Space, Learning to Live Without Gravity, Goldfinger, Ronko Spaceship Repair Manual, Aliens—Aliens?* Give me a break. I don't have time to sift through this mess.

Aha—the computer! Of course. Why should I look for the cartridge myself when this computer can probably do it for me? That looks like some sort of retrieval arm. Yeah, I'll bet this thing is made for finding cartridges. Smart thinking, Wilco. Now, if I can just figure out how to make it work.

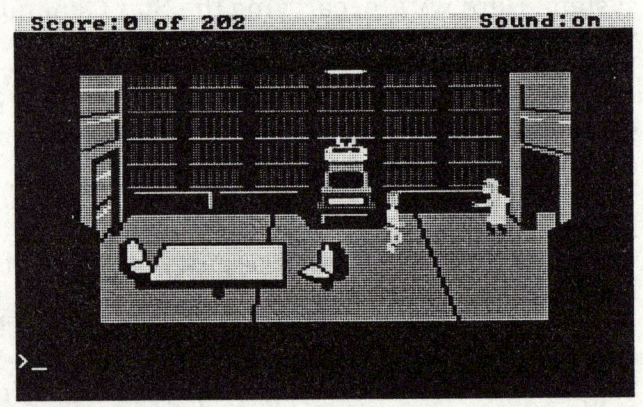

You aren't looking for books in this library. Hope you listened to what the professor said. Observing a computer screen is always a prudent thing to do.

Great! I've got the cartridge. Now what? There's no place to insert it. How am I supposed to view this thing? Well that's typical—a library full of cartridges and no way to view them. I might as well get out of here. Oops! Better not go the way I came in. Wouldn't want to run into another one of those Sarien characters. I'll just slip out the back way here. Maybe the professor wasn't the last to go. Maybe I'll find someone else alive. Maybe . . .

"Oh, no! Willie! Andy! My buddies! We used to play checkers every night in the third floor custodian closet—and now you're dead! We'll never play checkers again."

But I can't let my emotions get the best of me. I have to be strong, think fast; otherwise, I'm doomed like the rest of the crew. I need to protect myself. Surely one of my buddies carried a ray gun or something—whatever they're called these days. "OK, Willie, forgive me, but I have to search your uniform. I'm sorry you're dead, but I would let you search me if I were dead."

Nothing there. "OK, Andy, it's your turn now. Forgive me, buddy, this is hard for me to do." Nothing again except a keycard—just like the one I used at my old job back home. "I'll keep it for sentimental reasons, Andy, since it belonged to you. No wonder you guys didn't stand a chance—you weren't armed. But then, of course you weren't armed—no one on board carries weapons."

Why didn't I think of that to start with? The *Arcada* is an unarmed ship. No intercom, no weapons—what a piece of cake for the Sariens. Whose brilliant idea was it for us to play host ship to the Star Generator anyway?

It looks like I can forget about protecting myself. The only thing I can do is hide. At least I'm good at that, but I'm used to hiding from my boss, not trigger-happy Sariens. My only alternative is to find a way off this ship. Better take the elevator down to the next level. Seems to me I remember cleaning out a room used for flight preparation. Maybe that's my ticket out of here.

Arcada Lower Level

OK, now that I'm here on the second level, I've got to get up enough nerve to leave this elevator. Come on, Wilco, you can do it. You can do it. Just poke your head out the door and see if anyone's there. You can do it. OK, here goes.

Whew! The coast is clear. I'll just keep my cool and walk quietly. . . . What's that I hear? Footsteps again? Y-y-y-yikes! I'd better dash into that elevator again! Whew! That was close! The coast is clear up here, too, but I need to be down there, not up here. I'll try it again. Maybe that jerk will be gone this time.

OK, I'm back on level 2 again. I'll just calmly step out of this elevator and into the hallway. Whew! The coast is clear again . . . and I don't hear any footsteps. Nope, there's no one else here now; it's just me and my. . . . "Boss. Oh, no. Boss! It can't be you! Oh, dear. The Sariens got to you, too. Oh, gee. I'm sorry, Boss! I'm sorry for all those times I played sick but really goofed off, and for all those hours I spent sleeping in the closet when I was supposed to be working, and for all those terrible names I called you! I hope you can forgive me, wherever you are!"

All right, get control of yourself, Wilco. I've got to get the heck off this ship! This is no time to start feeling guilty about being the *only* person alive—just because I was sleeping in a closet when I should have been working. Now, what's that burned-out hole in the wall over there? I'd better check this out. I have a feeling I know where this is leading.

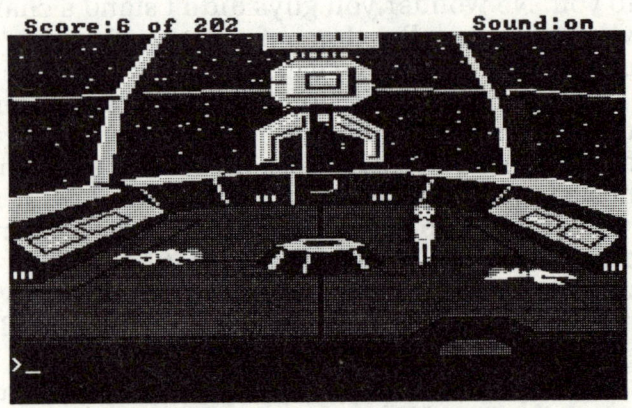

The only thing you'll find in this room is what's missing.

I was right. This is the Star Generator room—at least it used to be. The Sariens *did* steal the Generator, and there's no telling what they're going to do with it. I know one thing for sure—hanging around this place isn't getting me any-where. I've got to make my way to that flight preparation room and bid this bird adieu.

It seems to me I remember passing through a control center. Gosh, it's so hard to remember. I only cleaned that room once—every time I went near the place I was told to find another floor to mop.

Control Center

This must be the control center here. I guess that's what you call it. Looks like it controls something, with all these buttons and knobs and these big pulsating domes. How am I supposed to know what all these buttons are for? I'm no engineer. What's this? Open Bay Doors and Close Bay Doors. What bay doors? I don't see any bay doors.

Wait a minute—*bay* doors! Maybe there's a Galileo space shuttle in the bay—nah, not on this ship. Still, if I can find the bay, there's bound to be some sort of spacecraft I can use to escape. How the heck are you supposed to know whether or not the bay doors are open or closed? "Open the bay doors, Hal." No that didn't work. Guess I'll have to take a chance and see what happens. I'll just push this button.

Hey, neat! I didn't know there was a window here. Look at that view! Oh, gee. Come off it, Wilco. What's gotten into me? I'm standing on a time bomb whose time is about to run out, the entire crew is dead, the entire ship is infested with tacky red-suited aliens after my hide, and I'm standing here admiring the view! I've got to stop this nonsense and *think* like the swashbuckling hero I'm supposed to be!

OK, that's it. Think straight. What would a nonjanitor do in this situation? I think I've managed to open the bay doors—at least I hope I have—and I think I'm headed in the right direction. I'll keep going and pray that no Sariens are in the area. I see an elevator door over there. Yes, that must be the one leading to the flight preparation room I cleaned.

That's strange. The door should just slide open like any other elevator door. Now what's the matter? Don't tell me another Ronko gadget has gone haywire—this time something so simple as a door! Oh, wait a minute. There's a slot by the door. Maybe this is a private, exclusive, security-clearance-only entrance, which means one thing—I have to have a keycard to open the elevator door.

The keycard I found on my buddy Andy! Oh, please. Let it work! Please, please, please. . . . Yes! All right! I'm in! Thank you, Andy. You really were a true buddy!

Flight Preparation Room

Well, this is it. This is that flight prep room I remember so well. In fact, if I recall correctly, it wasn't even very dirty. I do remember these two closets, of course. Seems to me, one of them made a pretty good siesta room. Better check to see if there's anything important in there. Never know what I may need.

What's this gadget? Another Ronko deal? Looks like some sort of translator device. What the heck, if I have to plead for my life, at least they'll understand me. Hey, maybe I can even get a job on another ship or another planet! I'm not too proud to clean up after aliens, as long as they aren't Sarien scum. And now for the other closet. A spacesuit. Hmmm. Will I or won't I need the spacesuit? Any escape vehicle's bound to be pressurized.

Well, I see that the bay doors are indeed open now. These doors here are bound to lead to the bay. Hopefully, I'll find the right button on that control panel over there. What a pain to have to walk all the way around the room to get to it. OK, where's the button, where's the button? Gravity Control, Air Pressure, Atmosphere, Transporter Control, Airlock— Airlock! That must be it. Of course, one always passes through an airlock before entering the main ship. Thank goodness for that *Arcada* model builder the boss gave me for Christmas last year. Well, here goes.

Pod Bay

Great! That did it. Gravity seems to be normal, but boy, I can really feel the pressure change. I didn't even *think* about the consequences of those bay doors being open. Oh, no. There's no escape vehicle here. Someone must have escaped before me—either that or someone left the ship before the Sariens invaded. Maybe there's a way I can communicate with the spacecraft. Then, whoever's in it can come back and rescue me! It's worth a try. Oh, please be a communication device or something on the control panel!

Drat! Nothing! Just a button that says Platform. What in the world is that for? Oh, well. What do I have to lose? Might as well push it and see what happens. I've no way of escaping here now. The ship's going to go up like a nova, and I'll be sitting right on top of the fireball, my hair all singed. . . . Wait a minute! What's this? I don't believe it! It's a bird! It's a plane! No, it's a space pod! Holy cow! I may make it after all! If I can only figure out how to get in this thing. It's not very big. Gee, there are only so many places to hide a door. Aha! Found it. Now we're in business.

Aboard the Escape Pod

OK, I'm buckled up, and the door is closed. Safety first. Let's see; there are three buttons here—Power, Don't Touch, and Auto Nav—and a throttle.

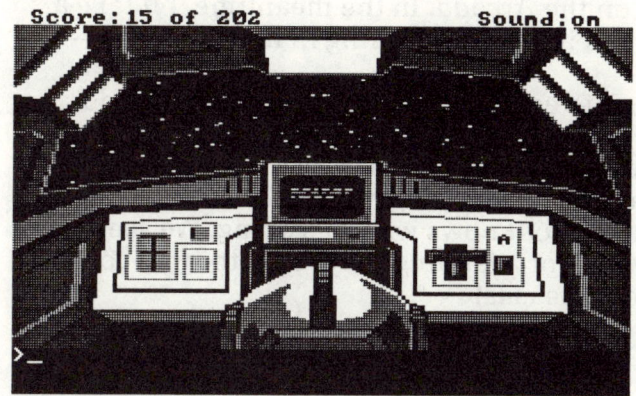

Score: 15 of 202   Sound: on

You're a doctor . . . I mean janitor, not an engineer! Let the escape pod fly itself.

Better play it safe and push Power. Well, that certainly did something. The engines are going. Great! Now I'll pull the throttle. Hey, I'm moving! This is unbelievable! Me, Roger Wilco, the janitor. I'm actually going to escape! Now that I'm out in space, I wonder where I should go. And I can't help but wonder what would happen if I were to push that other button.

Score: 30 of 202   Sound: on

No, you aren't in Kansas, Toto, and you sure aren't in Ulence Flats!

It never hurts to see what the control panel has to say. Hmmm, I've never heard of Kerona. Apparently, neither has the pod's computer. Oh, well. It's out of my hands now. The pod's going to go where it wants to, and there's not a darn thing I can do about it. Wherever I go, it's got to be better than blowing up on the *Arcada*. In the meantime, I'll take a look around and see if there's anything in here I can use—like food!

Nope. Nothing to eat, nothing to drink, nothing to do anything with. I guess that doesn't matter anyway; I'm sure I'll find something to eat on . . . what was that name again, Kerona? And that must be it straight ahead.

I hope this autopilot knows what it's doing. We seem to be going awfully fast for a landing. Hey! This thing is going to crash! Whooooooooaaaaaaa!

Welcome to Kerona . . . ooh.

Game Points Earned

Action	Points
Escaping the *Arcada*	15
Getting the Astral Body cartridge	5
Getting the keycard from a crew member	1
Getting the translator gadget	2
Lifting the pod on its platform	1
Listening to the professor	2
Opening an elevator with the keycard	2
Opening the pod bay doors	2
Using Auto Nav in the space pod	2
Wearing the spacesuit	2

Object Locations

Object	Location
Astral Body cartridge	Cartridge library
Keycard	Dead crew member
Spacesuit	Closet in flight room
Translator gadget	Closet in flight room

The Planet Kerona

Mission: Find your way out of the desert and into the town of Ulence Flats. Learn the contents of the Astral Body cartridge.
Total Points: 74

The Keronian Desert

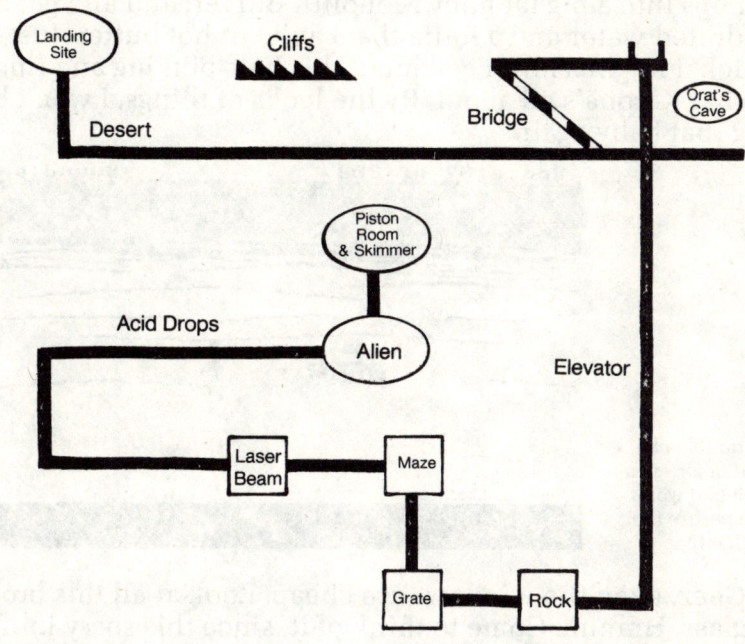

Inventory

body part
dehydrated water
glass
plant
rock
skimmer
Xenon army knife

Keronian Desert

Ouch! Some landing. Stupid pod—should have flown it myself. I could've landed better than that with my eyes closed,
and I'm not even a pilot.

What's this on the floor? *Survival Kit.* Oh, boy. I sure
hope this is what I think it is. . . . *What?* No food! It's just a
lousy Xenon army knife—cheapest brand you can buy. . . .
That figures—and dehydrated water. Great. I'm dying to sink
my chops into a big fat juicy Monolith Burger, and all I get is
dehydrated water and a knife that can't cut hot butter. Just
my luck. I might as well get out of this half-pint jug and find
out what Kerona's all about. By the looks of things, I won't be
flying that baby again.

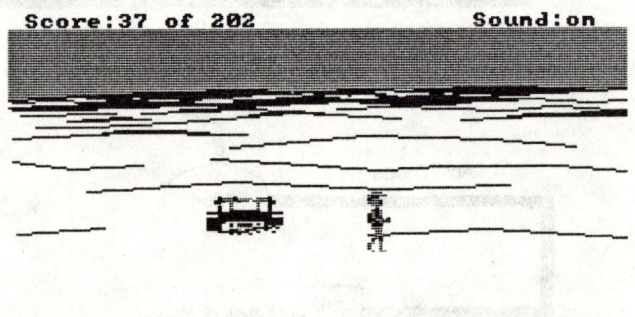

*Better double-check
the pod before you
bid adieu. It won't
hurt to peruse the
ground, either.*

Geez, even the windows are cheap. Look at all this broken glass. Hmmm. Come to think of it, since this sorry knife
won't do me any good, I should probably take a piece of this
glass. Who knows when it might come in handy.

Now, which way should I go? This is a pretty darn big
desert. I see some cliffs over there in the distance. I think I'll
investigate them. Besides, I don't see that I have much choice
other than miles and miles of sand. East it is. Looks like the
cliffs span northward. I think I'll head a bit northeast on second thought.

Hey, w-w-what was that? I *know* I saw something pop
out of the sand. Boy, my heart's beating so fast I can hear it in
my ears. I don't know anything about this place. There's no
telling what I'm going to run into. I don't know if I can take
this unknown stuff. Who knows what danger lurks beneath

the desert! What if I escaped the _Arcada_ only to be swallowed up by a sand monster?! I'd rather be taken by the Sariens!

OK, Wilco, just hold on to your sanity. I can do it. I'll just walk very calmly to those cliffs, and I'll watch every step I take. I can do it. At least I think I can. . . . I think I can. . . . I think I can. . . . Almost there . . . I think I can. . . . I think I can. . . . I _know_ I can. . . . I _know_ I can. . . . I KNOW I can. I can! I did! I'm here! I'm the little janitor that could! And I did!

Now to find the way to somewhere—anywhere. There must be a civilization around here. I see a bridge over there. Maybe that leads to something. From the looks of it, I'd say quite a few feet have crossed it—it looks like it's about to come tumbling down. The entrance must be south of here, but not too far south—these cliffs don't go on forever.

Why cross the bridge, you ask? Because it's there.

Desert Cliffs

What's this? A hole in the cliff. I wonder, why would there be a hole in the wall of the cliff right here? Maybe I should stick my head in and. . . . What's that noise? What's that _thing_? A flying disk? It's sprouting legs. It's a-a-a mechanical SPIDER! And it's got a big fat eye that keeps looking around—like it's looking for _me_! Ahhhhhhh! I _hate_ spiders! I'm afraid of spiders! I've got to get out of here!

I've . . . _(puff)_ . . . got to find the entrance to that . . . _(puff)_ . . . bridge! It must be . . . _(puff)_ . . . a little south and then . . . _(puff)_ . . . east. I've . . . _(puff)_ . . . got to find it. Oh, yes. There it is. Save me, bridge. Save me. I hope that spider is afraid of heights.

31

It's the nature of man to kill spiders. It's the nature of this spider to kill you—or maybe someone else, given the opportunity.

Whew! This is getting to be a little too much. Thank goodness that thing can't come up here—at least, I don't think it can. It must be programmed for the ground or something. But what if it isn't? What if it's only a matter of time before that spider sees me up here and heads for the bridge? If I could only figure out a way to get rid of it. Just look at it—crawling around the ground, looking for *me*. If it weren't so big, I'd step on it and crush it to death.

Wait a minute—a rock! Maybe I can pry this rock loose and roll it off the bridge when the spider comes by. Oh, heck. It's worth a try, but I'll have to make sure I push the rock at just the right time.

Be patient, Wilco; just be patient. Come on spider; come on. That's it. Just a little closer. That's it. . . . NOW! *Got it!* I did it! I can't believe it. Yeah! I'm getting pretty good at this, if I say so myself.

Now where to? I'm dying of thirst—I need a drink of water. This dehydrated stuff looks pretty bad. . . . Ugh. It *is* bad—no taste at all. I saw some plants down there on the surface that might be edible—I'm really starving. But first, I think I'll check out that odd-looking pair of rocks over there. Looks almost like it used to be an arch or something. An entranceway maybe? But why out here in the middle of nowhere—and an entranceway to what?

Hmmm. This is a strange-looking pair. But what does it me—Whoooooaaaa!

Underground Cavern

What happened?! I'm in an elevator—and heading straight down—and I don't see any buttons to push to stop this thing! Where the—Oops, better not say that word. It might be bad luck. I don't want to find myself facing pitchforks and point-ed tails.

Obviously, this is no ordinary elevator. It cuts through solid rock. Who in the world would build such a thing, and why? Of course, I have to remember that I'm on a strange planet. I can't expect anything to be normal.

Oh, great. I started nowhere, and now I've ended up no-where. This is nothing but a bunch of caves—and ones that probably lead to nowhere again. What kind of rock is this anyway? Doesn't look like any I've ever seen before. There's a smaller piece over there. Let's see; what's this thing made of? Not the same as the cliffs—that's for sure—but obviously indigenous to this planet. Wow, I'm sounding smart here.

Saaay, where's the light in here coming from? . . . Er . . . let me rephrase that. Ahem. I am curious as to the source of illumination occurring in these subterranean conduits. Yep, I'm sounding smart all right.

OK, let's get on with it. I'll just tuck this little specimen in my pocket—maybe I'll have a chance to study it later. At least there *is* light in here, wherever the source. Looks like there's only one way I can go. No telling what lies around the corner.

What in the world? A grate of some sort. Maybe that's an air shaft. There has to be a source of air in. . . . Whoooaaah! What was that?! Oh, my gosh. It has long slimy legs, and it's trying to grab me! That's no air shaft—that's a cage for some creepy, tentacled, slimy-legged *thing!* I don't know how much more of this I can take! I've got to get out of here. I'm starting to feel like the gingerbread man. If only I could run as fast as he can.

There has to be a way around this thing. Maybe if I squeeze by—as close to the rock wall as possible—its tenta-cles won't be able to reach me. I've got to try. There's no al-ternative. Boy, am I tired of these no-alternative situations. OK, here goes nothing. . . . D-d-don't t-t-touch m-m-me, p-p-please! Y-y-y-y-y-yikes! Oh! I made it! Whew! Now to just get the heck out of here! But how?

That looks like a door over there, but it's closed. How do I open it? I have to think fast. That geyser—what is a geyser doing way down here? It's giving off so much steam—this place is as hot as Hades. Well, this is just dandy. I risk my life to cross that grate only to die of heatstroke. Hey, maybe I can plug up the geyser with this rock I found. It's bound to help some, even if it doesn't completely shut it off.

OK, here goes. . . . Hey! I don't believe it! That *is* a door, and now it's open! All right, I'll just make my merry way through there—and away from this sauna.

What kind of place is this anyway, and what is that stuff dripping? I'll have to take a closer look. Ohhhh, I see. It's *acid.* Don't think I'll be taking a swim in *that* pond anytime soon. No wonder this place is so cavernous—acid must be a natural substance on this planet. And it's like a maze down here. How am I going to find my way out? That ledge overhead goes somewhere. The problem is finding my way to the ledge. There has to be an access path somewhere around here. OK, if that's what it takes, I'll walk through every opening in here until I find my way out.

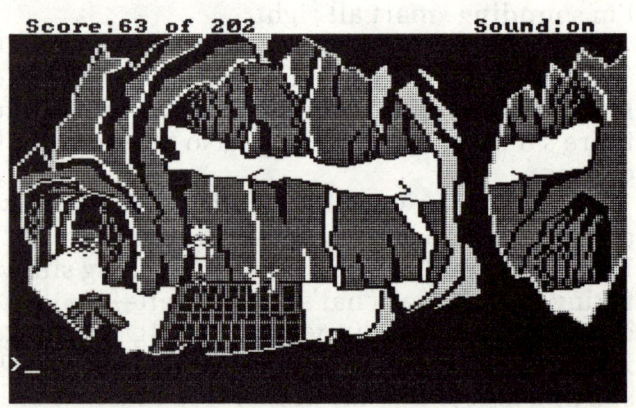

Better watch your step! Oh, and take it slow when you encounter "acid rain."

Found it. I knew there was a way out of that rat's maze. Oh, no. Now what? Not another obstacle in my path, pleeease. For some strange reason I think the best course of action is *not* to walk through that beam—not unless I want to be someone's fried fish for dinner. Think, Wilco, think! If I had my mop and pail, I could throw water on it. Nah, that's no good. There's no water in here except this bottle of dehydrated water. Let's see; what else have I collected here—knife, cartridge, translator gadget, glass. I don't see any wires to cut on that contraption, not that this so-called knife could do the trick anyway. But wait a minute! This piece of glass could! Maybe if I try cutting away one of these beam emitters. . . . Hey! It worked! This is better than cutting. I didn't need anything sharp after all, just something reflective. Boy, am I glad I picked up that piece of glass!

Now I see the way to that overhead ledge. I knew there was an access path somewhere. I'll just follow the yellow dirt path until I reach the land of . . . uh-oh. How am I going to make my way through this mess? I swear, I think I'm out of the woods, and then every darn time I turn around there's something else in my way. Slowly dripping acid. I can't let that stuff touch me. I'd better be careful and take my time through here. If those drops can burn through rock, no telling what they'll do to me. OK, one step at a time. Just wait until the next drop and. . . . I did it! Once more, Roger Wilco beats the odds. *Now* maybe I can find my way out of here!

Gee, I was expecting things to get brighter, but all that natural light has gone. Everything's getting darker. Maybe this translator gadget has a light on it. At least that would be better than nothing. OK, I've turned it on. Great, one little red light. How's that supposed to help? This is no good—I still can't see anything.

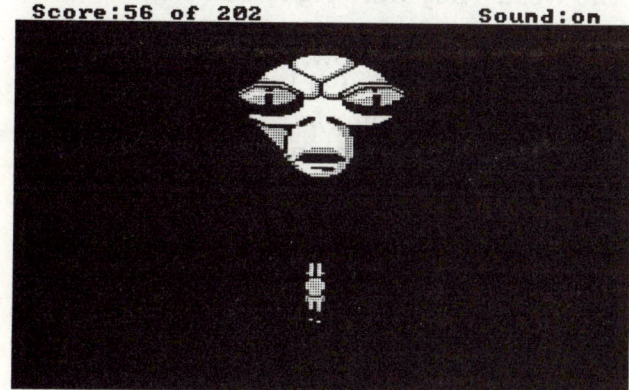

```
Score:56 of 202                    Sound:on
```

This fellow has a terrible accent. Can't understand a word he says.

Wha—huh? I can *definitely* see something now! W-w-what's that?! It doesn't look very friendly! In fact, it looks quite monstrous! And it looks like I'm up a creek!

What was that? That thing said something to me! "Uh, excuse me, sir, but I was wondering if you could help me. . . . Sir?" It's not listening to me. It's just talking. "What? I have to find a *what* to prove myself? What is an Orat, or rather, *who* is Orat? I have to bring back proof that I've done away with Orat? Hey, mister, I don't know anything about any guy named Orat, and besides, my knife is really dull, and. . . . Hey!"

Desert Cliffs

"What's happening? What?! . . ." I'm back at the top of that cliff again. Oh, gee. This isn't so great. I thought I was on my way out of here. He *did* promise me I could leave if I bring back something to prove I've gotten rid of an Orat. Well, how hard can that be? This is getting ridiculous, though. I'm back to square one, and I haven't had anything to eat in hours—just this disgusting powdered water.

Before I find any Orat, I'm ready to pig out on some veggies. Now where were those plants I saw? Let's see—down the bridge and to the west. They were right near the bridge as a matter of fact.

There they are! I'm in business now. OK, plant, you may not be a yummy Monolith Burger, but at least you're food—I hope. OK, here goes . . . (*chomp chomp*). . . . Yuck! Ugh! Blah! This thing tastes worse than Sarien onion stew! I've got to force it down, though. I'm getting too weak from starvation. Under any other circumstances, I'd be trashing this weed.

Now that I'm about to throw up, I guess it's time to find ol' Orat, whoever he is. And I wonder where he lives! Maybe in a cave! I don't recall seeing any caves on this side of the bridge, only that hole. Maybe I should go back and check it out. Nah, it's too far away, and I'm anxious to get this show on the road and get OUT of here. I'll have to look on the other side of the bridge. Maybe there's a cave in the wall across from it. I'll just retrace my footsteps back to the bridge path.

Now, I should probably explore that entire wall that runs north. OK . . . I don't see anything here. . . . Nothing here . . . nothing here . . . noth. . . . Hey! Here it is. This must be it. It definitely looks like the mouth of a cave to me. OK,

Wilco, what are you waiting for? Just rush in there as fast as you can, and. . . . Wait a minute.

I don't have a plan. I don't even have a weapon. What in the world am I going to do? I'll just go slow and check it out *before* charging in. Maybe Orat's a little elf or something— oh, please be a *little* elf!

OK, here goes. M-m-my knees are kn-kn-knocking. I hope he doesn't h-hear them—or m-my t-t-t-teeth chatter-ing. I'll j-just take a deep breath and go slowly. I don't see anything yet—just a big rock. Shoooh, it stinks in here! What does this guy do for a living, collect garbage and keep it? What was that?! I see him . . . it! Oh, gee whiz. This isn't fair. He's BIG. Too bad I knocked off my eight-legged friend back there. I'd like to see those two get together. Oh, no! He saw me! He's coming over here! I don't have time to escape! What am I going to do?!

"Uh, hi there. Nice day, huh? Your place here really smells bad, but tell you what I'm going to do. This is your lucky day, you know that? You see, I just happen to be a jani-tor, and uh . . . er . . . hey! Want a drink of water? Well, take one ANYWAY! Here!"

Huh? He took it. I mean he drank it. I mean he ate the *whole* thing! Uh-oh. I think I know what's going to . . . hap-pen . . . NEXT! Wow! That's what I call going out with a bang. That fellow wasn't too friendly—can't say I'm sorry to see him go. That was a bit close for comfort, though. If I hadn't had that water with me. . . . I don't even want to *think* about what would have happened!

Now I need proof that I killed him—but how? He was blown to smithereens when he ate that water jug. Maybe I can find a piece of shredded clothing or something on the floor. What's that? Looks like a. . . . Eeeewww, yuck! It's an Orat part—of what, I don't know. Oooo, this is gross. It's sticky and gooey. I need a plastic bag—too bad I don't have one. Oh, well. I have to take it. The smell's about to knock me out, but at least I don't have far to go. Back to the alien hideaway now. Let's see. . . .

Over the bridge and through the rocks
Down the elevator I go.
I'll squirm past the grate,
Cross the laser-beam gate,

And through acid rain I'll tiptoe, oh!
Over the bridge and through the rocks
To the alien's cave I go. . . .

Here again, finally. "Uh, Mr. Alien Head. . . . Yes, sir. I
have the part here you requested to prove that Orat no longer exists—at least not in living form—thanks to me. Yes, I'll
drop it right here in front of you. By the way, I just have to
say that I had to go into battle unarmed, and I don't for one
minute appreciate. . . . Hey, are you listening to me?"

This guy likes to hear his own voice. He hasn't heard a
word I've said. What's that? Another opening? At least he
isn't sending me to the surface again. Maybe this is the *real*
way out.

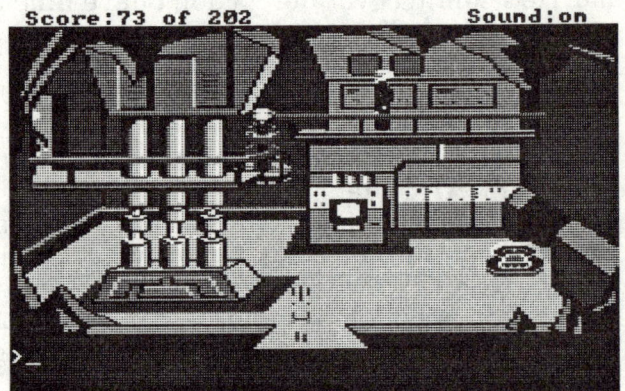

Score:73 of 202   Sound:on

*These folks are real
Wizard of Oz fans.
They'll help you
find your way back
to Kansas.*

Keronian Piston Room

What the. . . . Hey, you mean that big alien head wasn't even
real? It was just a projection or something. What kind of people are these anyway? You could never accuse them of being
short-handed. What's the point of a piston room beneath the
surface of a desert and cliffs? Oh, well. I guess mine isn't to
question why when they're nice enough to point me in the
right direction—and with transportation to boot.

"Thanks very much, guys. Catch you later. I'll just be on
my way now. . . . Saaaay, that looks like a computer over
there." I wonder, maybe I can finally learn the contents of
this Astral Body cartridge. Heck, it's worth a try. Let's see;
the directions say, "Insert cartridge in slot, magnetic tape

side down, as shown." OK, here goes. Probably some B movie anyway. . . . Oh, maybe not.

I knew it! The Sariens are planning to take over the universe with the Star Generator! But this tape says the thing can be activated to self-destruct. Now I know I *have* to find it. After all, what do I have to lose if the alternative is living in a Sarien-run universe? No way. I'll die before I clean up for that Sarien scum. But I have to remember this code—I can't forget it. Darn, I was never good with numbers—can't even remember my telephone number. I will *make* myself remember. Let's see—*six, eight, five, eight*. Those letters are s,e,f,e. Sefe? I can't remember that!

There has to be a way. I know, I'll make up a phrase. Let's see—*Sanitarian Equality For-Ever*. That's it! I can remember that! I've been pushing equal rights for years now! *Sanitarian Equality For-Ever!* Way to go—I'm really clever!

I'll just take this cartridge back, and now to get going—finally! Looks like there's only one way to get in this skimmer—I think that's what he called it. And I'm going to. . . . Where did he say, Ulence Flats? OK, turn the key, and I'm out of here!

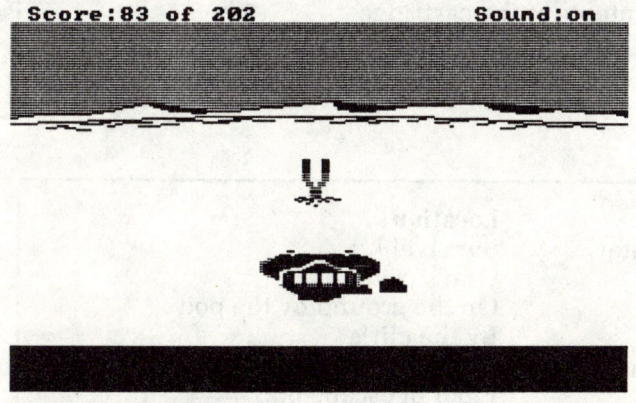

Keep your eyes straight ahead and take your time—or you'll end up with your head in the sand.

Aboard the Skimmer

Well, this thing sure likes to take its time. He said the town was just over the horizon, so I don't want to lose sight of that. Hey! What's all this junk flying at me?! No one said anything about having to dodge a bunch of rocks. Whooooaaaah! Yikes! Oooooowwwww! That was close! Here comes another one. . . . Ahh! I'm hit! Ouch! At least this little baby's still going. Just keep my eyes on the horizon. I can make. . . . Yoowwww! Hit again!

Oh, come on. There's not much farther to go. It's getting closer and closer. Maybe if I slow down some and just keep my eyes focused up ahead. Whew, that helps a little! I'm almost theeeeeeere! Another close one! Almost there . . . almost there. . . . Yeah! I'm there! I made it! Civilization at last!

Game Points Earned

Action	Points
Acquiring the skimmer	10
Blowing out the laser beam	5
Discovering the underground elevator	2
Finding the survival kit in the pod	2
Getting the piece of broken glass	3
Getting through the acid drops	3
Killing Orat	5
Killing the spider robot	5
Opening the cave door at the geyser	4
Retrieving the cartridge	5
Taking Orat's body part	2
Viewing the contents of the cartridge	5

Object Locations

Object	Location
Dehydrated water	Survival kit
Orat part	Orat's cave
Piece of glass	On the ground by the pod
Plant	By the cliffs
Rock specimen	By the underground elevator
Survival kit	Floor of escape pod
Xenon army knife	Survival kit

The Town of Ulence Flats

Mission: Find a way to leave Kerona and get to the Sarien spaceship.
Total Points: 43

Ulence Flats

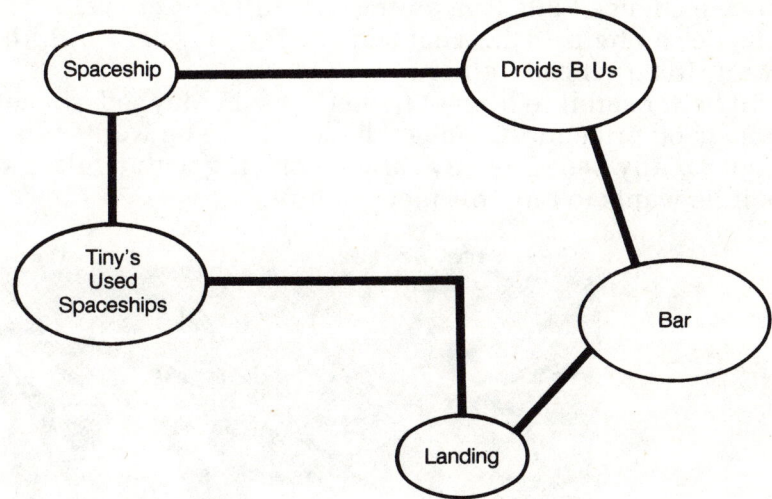

Inventory

beer
buckazoids
jetpack
pilot droid
skimmer key
spaceship

In Front of the Bar

So this is Ulence Flats. What a great-looking place, and it looks like there's a lot going on here! I really lucked up finding this town. Boy oh boy oh boy! Not only is there a bar here, but there should also be a decent spaceship somewhere—something easy to fly. I guess Kerona's not so bad after all.

Uh-oh, who's this scraggly-looking guy headed my way—the Ulence Flats welcoming committee? Oh, well. Maybe he can help. "Uh, what can I do for you, sir?" Ohhhh, he wants to buy the skimmer.

I'd better sell it to him—I need the cash. Maybe I can get a really good price for it. I mean, he looks like he wants this skimmer really bad! The guy can't be playing with a full deck if he wants to buy this piece of junk.

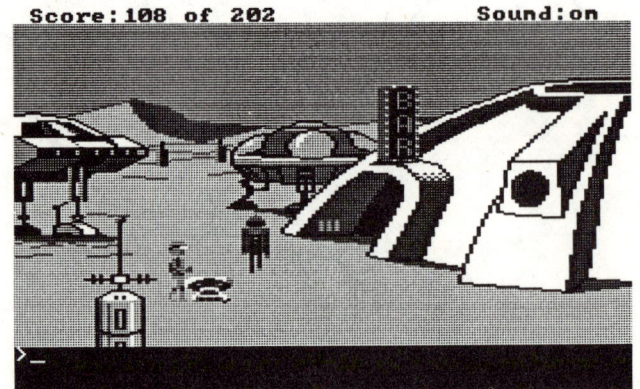

Score: 108 of 202   **Sound: on**

He knows a good skimmer when he sees one, but don't be in such a rush to sell it.

"You want to buy this sorry piece . . . uh, I mean, *savvy* piece of flying equipment? OK, what's your offer? . . . How much? You've got to be kidding! I'd be giving it away. No way. This is the ultimate skimming machine. I'm afraid you'll have to do better than that. . . . What? Hey! Don't you want to make another offer? . . . Fine! . . . No, YOU be that way!"

Great. I should have taken the first thing he offered. He must not be too dumb. As a matter of fact, he might be pretty clever. I'd better get the key out of the skimmer just to be safe. That clown's liable to come back here and fly away with my skimmer!

What now, though? I guess I could just walk around and see what's here—no point in going to the bar if I don't have any money; I'm sure they aren't handing out freebies. Hmmm, what's that place over there? *Tiny's Used Spaceships.* Oh, I'll bet he's selling the cream of the crop. Guess it won't hurt to find out.

Tiny's Place

"Well, I guess you're Tiny. . . . Yeah, thought so. Yeah, I'm sure you *do* want to make me a deal. I'm just browsing right now, so don't breathe down my neck, OK?" Geez, this stuff looks worse than the skimmer. These can't be for space travel, and every single one of them has obviously been totaled and given a new paint job—not even a good paint job. That figures. Nothing here I can use. . . . Saaay, that ship way over there looks pretty good. I wonder how much he wants for that one.

"Ahem. How much for that old thing back there? . . . HOW MUCH?! Who do I look like, Donald Trump? . . . Well, I'm NOT him. If I were, I wouldn't be buying anything from you, would I? . . . What? I have to have a droid to fly it? I don't have a droid. Do you see a droid here? Mister, do you know what I do for a living? I'm a janitor. I don't even have a droid to help *me* out, and unless you know of some recent developments in. . . . Huh? The Droids B Us store? Oh, and how much is *that* going to cost? Well, I certainly will ask them.

"By the way, I don't guess anyone around here could use a janitor of my esteemed reputation and talent. . . . Oh, I see. Droids perform the menial chores on this planet." Hmmmp. Gee thanks, Tiny, for making my day with that little ray of sunshine.

Oh, boy. I'm getting nowhere fast. Wilco, I think it's time to drown your sorrows. But with what? I don't even have the buckazoids to buy a glass of beer. Maybe someone at the Droids B Us store can help me. It's not too far from here. I'll just drag my feet in the dirt and kick a few stones as I make my way slowly and pathetically over there.

Outside the Droids B Us Store

Well . . . *(sniff)* . . . I'm here, but I know no one in there's going to help me . . . *(sniff)*. . . . N-nobody cares. My buddies are gone . . . *(sniff)*. . . . My boss is gone. . . . I'm broke. . . . I'm on a strange planet . . . *(sniff)* . . . in a strange town . . . *(sniff sniff)* . . . . Woe is me. I guess I'll just hang my tears out to dry. I guess I'll just c-cry me a . . . *(sniff)*. . . .

What's that on the ground behind Droids B Us? Hey, it's five buckazoids! I can't believe it! All right, maybe my luck is changing! Bartender, here I come! Roger Wilco is back!

Now to find my way back to the bar—I believe I'm going the back way. That must be it right there, that building with the window. What's that pile of sand there? Oh, well. THAT figures! You can't expect a machine to clean like a *real* janitor.

At the Skimmer

I see my skimmer's still there. Better check on it before I go into the bar. It looks like this thing's still my only means of transportation around here. Hmmm, here comes that scraggly guy again. Maybe he'll make another offer. Oh, I hope he does. This time I *will* sell it to him—first offer he makes. I'm not making the same mistake twice. I may not get enough to buy a spaceship, but I'll have a great time at the bar!

"Yes, sir. You're still interested, you say? Well, what's your offer this time? A jetpack, too? Well, I suppose that's a pretty good deal, although you realize you *are* the one who's making out like a thief? OK, sure. I accept your offer. Great, thanks. Here's the key. Enjoy!"

There's a sucker born every day. Yeah! Right now I feel rich. This still doesn't solve my problem of buying a spaceship, but what the heck. At least I can buy all the beer I want! Into the bar I go!

In the Bar

Hey, this place is really hopping, and what a great band! Wow, they're really good! Well, if I'm going to be stranded, it might as well be here. I'll bet these folks are all just passing through themselves. Maybe one of them can give me a ride to another planet—or anywhere!

The Sarien Encounter: Where No Janitor Has Gone Before □

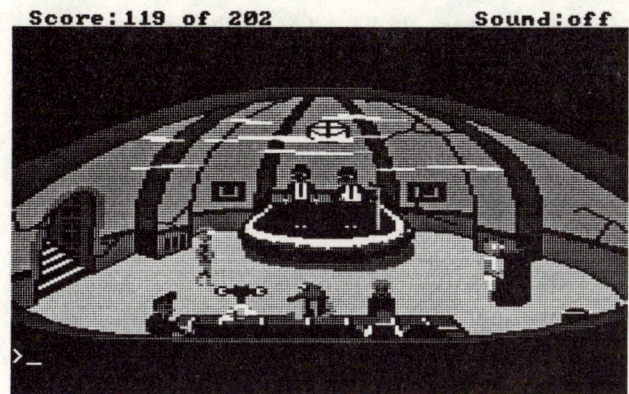

> _

So that's what hap-
pened to the Blues
Brothers!

"Hi there! My name's Roger. What's yours? Welcome to
the neighborhood, neighbor. . . . Excuse me, I said my name
is Roger and welcome to the neighborhood. . . ." Maybe he
doesn't speak English. Maybe he doesn't speak period. "Fine
with me, jerk. I'll just talk to someone else.

"Hi, how are you? . . . Excuse me, aren't those big flaps
on your head called ears? Doesn't anyone in here know how
to speak? I'm just trying to be friendly. My name is Wilco . . . Roger Wilco. I'd really like to get a ride out of town
with one of you, wherever you're going. I have buckazoids. . . . I can PAY if that's what you want! Is anybody even
going to say anything? OK, you lousy alien scum, be that
way! Your mother smells like Orat's cave!"

Might as well get a drink. "Bartender, I'd appreciate being served now. Great, I'll have a beer. Say, what's with these
creeps in here? Oh, I see. You're a busy little bee—too busy
with your silly little job here to talk to me, too. Well, fine.
Give me another beer and don't expect any tips. What is this
stuff, anyway, Keronian acid? Ha-ha." Jerk. I don't care if
you laugh or not—I thought it was funny. "Hey, bartender.
Send another one over here. Of course, I have the money.
What do you think I am? Never mind; don't answer that."

Oh, I see. They're too good to talk to me, but they don't
mind talking to each other. Who cares about their silly clique
anyway. Wait a minute; what was that he said? . . . Something about a sector? He said sector HH. I'll bet that's exactly
where I need to go to find the Sarien ship. I've got to remember HH. Let's see; how can I do that? This shouldn't be as
hard as memorizing numbers. How about . . . *Handi-Helpers,*

45

the name of those cleaning towels I use! Yeah, that's it—
Handi-Helpers. That's even easier to remember than *Sanitar-ian Equality For-Ever*.

Well, I think I've had enough of this sludge they call beer. I probably *shouldn't* have another one anyway. Oh, look. A slot machine! Boy, I love to play those things, espe-cially the ones at the Interstellar Fair. I can't resist—I'll have to give it a try. Besides, what do I have to lose? Don't have enough money for a spaceship, and the beer's lousy. Might as well give it my all—buckazoids, that is. OK, Lady Luck, I hope you're sitting on my shoulder now. . . .

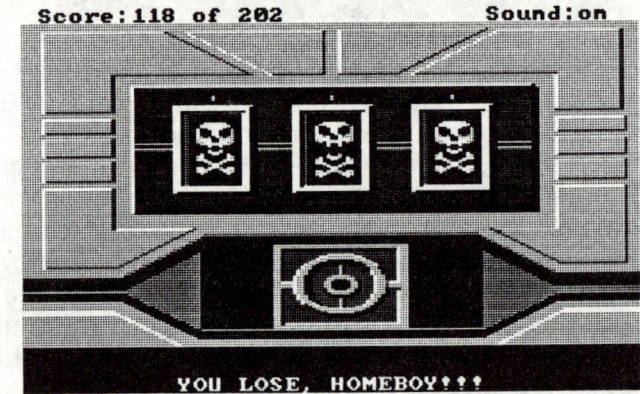

Looks like a ren-dezvous with death. Hope you saved your game.

Darn it! I've been playing this game for an hour, and I still don't have enough money to buy that spaceship. I keep saving it, but I still can't get anywhere. Something had better happen soon. OK, one more try, and then I'm going to give it up for a while. I never did check the price of those droids. Now might be a good time to do that. I have a feeling Lady Luck is getting ready to undergo a face lift, and I mean liter-ally. . . . Uh-oh, one skull, two skulls . . . y-y-y-yikes! I'm not waiting around for the third one to pop up. I'm nobody's fool! Suddenly, that pile of sand outside makes a lot of sense. I'm out of here!

Outside the Bar

Whew! I have a feeling I just narrowly escaped permanent dehydration! Oh, no. What now? Who's this guy coming over here? Well, maybe he can help.

"You have a spaceship you'll sell me cheap? Gosh. Well, gee. That sounds great! I'm glad to know there's someone decent in this town. How much do you want for it? . . . I have that much! Yes, sir. I sure do! And it doesn't need a droid? . . . Great! And you're sure it's a *spaceship* for *space* travel? Do you mind if I look at it first? Wow! That sounds like a great deal! I sure appreciate your being nice and helpful and all!

"Saaaay, wait a minute. How did you know I was looking for a spaceship? Oh, word travels fast, does it? Well, I'm not as stupid as I look, Bozo, so beat it! I'm not going anywhere with you. I bet you don't even have a spaceship."

I'm glad that character's gone—he's obviously on the take. Spaceship, my foot. I'm glad I didn't fall for that nonsense, or I'd be back at square one with the slot machine! Now, where was I? Oh, I need to buy a droid—hopefully, it won't be too expensive. I don't suppose there's another five buckazoids lying on the ground. OK, here goes. Maybe I can cut a real deal this time.

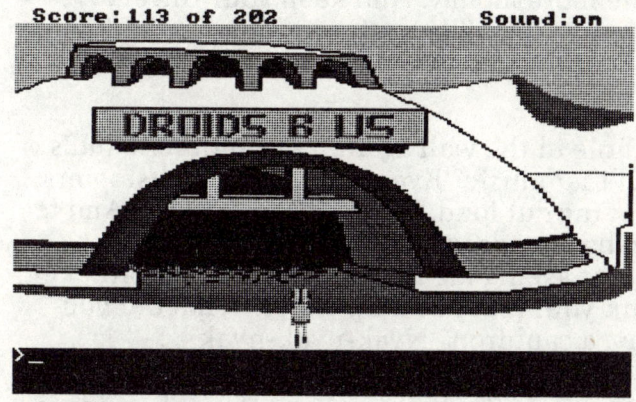

At least one droid "Should B Yours." Look for the fly-boy.

In the Droids B Us Store

Great, another sleazy salesman. "Hey, Bucko, why don't you let me browse a few minutes before going into your spiel? I'm not in the mood to hear it, OK?" Gee whiz, these droids look too big to pilot a small spaceship—and they're too darn expensive on top of that! Hey, those up there are on sale. Maybe I should have been a little nicer to the salesman. . . . Nah, they're all alike. He wouldn't give me the time of day anyway if he knew I was a janitor.

"How much for this one here, sir? And I'm not rich, so forget about overpricing. I can only pay so much, comprende? Oh, well. That sounds very reasonable. There's no catch to this, is there? OK, then. We have a deal." It's about time. I'm getting sick and tired of this town.

"OK, Fly-Boy, follow me. We're off to the slot machine for another hour of ripoff city. Well, why don't we chat on the way back to the bar? I realize you're only a droid—oh, I'm sorry—I realize you aren't human, but maybe you're the friendly type? The others around here don't know how to speak unless they want to sell something—slimy bunch of weasels. Well, you seem to be pretty friendly. Does that mean you talk?

"Oh, I see. You do talk. Yes, you talk all right. I get it, believe me. Yes, you *definitely* talk! Well, after what I've been through, talking too much is certainly better than not talking at all. I think we're going to get along just fine. Now, why don't you keep yourself busy out here while I go in and attempt to win some more money. And keep your fingers . . . uh, circuits crossed. Maybe I'll survive another round."

In the Bar

OK, back to this hole in the wall again. I see another band's in here. They call that music? Even the same old customers are in here, for crying out loud. What a boring group. And to think I wanted to be friends with these clowns! "Hey, bartender! Don't bother asking me if I want a drink. I'd rather eat dirt than drink your lousy swamp water. What did you brew it in anyway, a cauldron? Nyak-nyak-nyak. . . ."

Back to the slot machine for another zillion hours of exciting play. . . . "Finally! I finally have enough buckazoids to buy that spaceship. And as for you, you rigged-up piece of garbage, I hope no one EVER plays your stupid game again. Hey, bartender. All three of your bands stink. You call that music? Don't worry about seeing my face again—I'm blowing this cruddy joint for good! Adios, los jerkos.

"Come on, Fly-Boy. We'd better hightail it over to Tiny's! The bartender and his patrons weren't looking at me too nicely when I left. Can't imagine what the problem is."

Tiny's Place

"OK, Tiny. I'm ready to deal with you, but not one buckazoid higher than the price you quoted me the last time I was here. What do you mean you don't remember me? I'm the guy who was asking you about that ship over there. You said I needed a droid to fly it, so I bought one. Now I have the bucks to buy the ship, so is it a deal? Great. Here's your money. Thanks, and have a nice day."

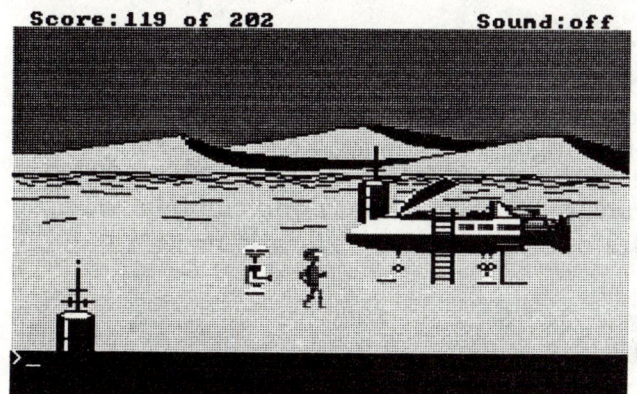

You're still no pilot. Let someone else do the honors—just make sure he's loaded.

The Space Shuttle

"OK, Fly-Boy, I hope you know what you're doing. I'll climb up into the seat here, and let's see. What do I need to do next? OK, load her up! Great, we're on our way." I can't believe I'm finally leaving this wretched planet. I'm so happy I could cry. And I have my little droid friend, too! "What's that, Fly-Boy? Sector? Let's see. Where did I hear about a sector? OK, hold your horses—I'm thinking. Oh, I remember! It was in the bar. Yeah, that's easy. The sector is *Handi-Helpers* . . . uh, I mean, HH. Just got a little mixed up there!"

Wow! We're really on our way. I wish I could relax and enjoy the moment, but all I can think about is the Star Generator in the hands of the Sariens. Sector HH is where the Sarien ship will be, if I overheard correctly. I'm not looking forward to this, but I may be the only hope of the universe!

How did I get from cleaning toilets to this? I think I'd rather carry the burden of too many mops and brooms and cleaning fluids than the weight of saving the entire uni— "Whoa, Fly-Boy! What's all that commotion out there? We're in an asteroid storm? Did your readings indicate we'd be passing through a storm?" That's very odd.

"Is there anything we can do?! Do you think the ship will hold out?" Oh, I sure hope so. Please don't let me have come this far only to be blown away by a bunch of flying rocks!

"They're gone! Oh, Fly-Boy, we made it—thanks to you and your flying expertise! What would I have done with . . . with. . . . Oh, my. What in heaven's name? . . . Is that the Sarien spaceship?"

It is, indeed. That's the *Deltaur*. Suddenly, I'm not feeling so well. Where are the good old days when I could whistle while I worked? Maybe this wasn't such a good idea after all. I mean, this is going to be more than a little reconnaissance mission. This involves high-level undercover work and sabotage! I never said I was James Bond! Somehow, I don't think my training as a janitor is going to be very helpful on this ship.

Game Points Earned

Action	Points
Eavesdropping on the bar patrons	5
Getting the jetpack	5
Leaving Kerona	25
Making it to Ulence Flats	25
Purchasing Fly-Boy	4
Purchasing the correct spaceship	4

Object Locations

Object	Location
Beer	In the bar
Buckazoids	On the ground and from the slot machine
Fly-Boy, the droid	Droids B Us store
Jetpack	Scraggly old man
Skimmer key	In the skimmer
Spaceship	North of Tiny's store

The Spaceship *Deltaur*

Mission: Find and detonate the Star Generator.
Total Points: 50

The Spaceship *Deltaur*

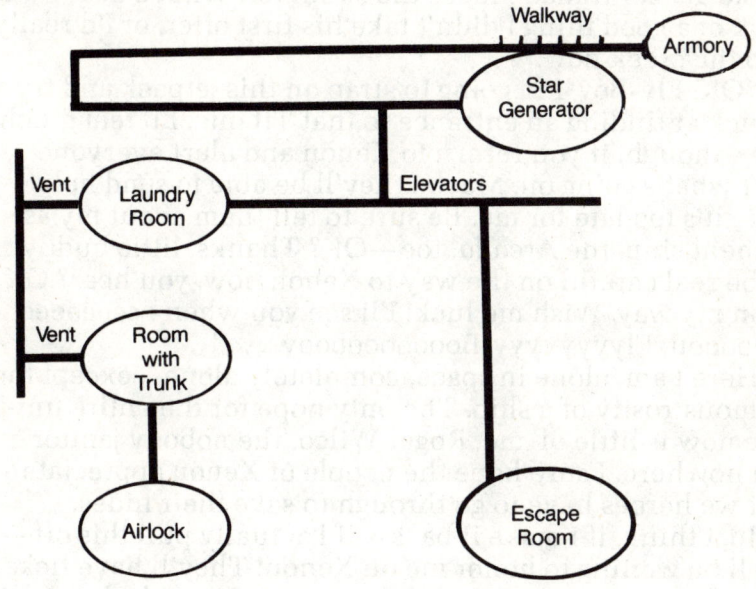

Inventory

escape ship
gas grenades
identification card
pulseray
remote control

Outside the *Deltaur*

Well, first things first—and first I have to figure out how to get aboard that ship without being seen, which means I have to sneak in somehow. Of course, that's the $64,000 question. Once again, I'm forced to think like a nonjanitor.

I've got it—the jetpack! I'll fly aboard using the jetpack that old geezer traded me for the skimmer! What a deal—and a heck of a good thing I didn't take his first offer, or I'd really be up the creek now.

"OK, Fly-Boy, I'm going to strap on this jetpack and try my luck at finding an entrance to that Titanic. I'll feel much better, though, if you return to Xenon and alert everyone about what's going on. Maybe they'll be able to send help before it's too late for me. Be sure to tell them about my assignment ship, the *Arcada*, too—OK? Thanks, little buddy. You be real careful on the way to Xenon now, you hear? OK, I'm on my way. Wish me luck! I'll see you when I seeeeeee yoooooooou, Flyyyyyyyy-Booooooooooy. . . ."

Here I am, alone in space, completely alone—except for this monstrosity of a ship. The only hope for the entire universe now is little ol' me, Roger Wilco, the nobody janitor from nowhere. I sure hope the people of Xenon appreciate what we heroes have to go through to save their hides.

Just think, if I make it back—if I actually pull this off—they'll be waiting to honor me on Xenon! They'll have ticker-tape parades in my honor with big red-white-and-blue cakes with my name on them. My face will be all over television—from sea to shining sea, from Neptune to Pluto, from the Milky Way to Andromeda! They'll erect statues of me in every city! I'll be a national symbol of courage and bravery with my name ringing freedom throughout the land—throughout the universe! I'll be a legend in my own mind. . . . I mean, time! It's worth it! I can do it! I'll go for it! Uh, but first I have to figure out how to get inside the *Deltaur*.

Let's see. This looks like a doorway—probably leads into an airlock. If I could only find the lever to open it. . . . Ah, there it is! Thought I'd never find it, but then what did I expect? Sariens are as cheap as they come. Sariens are so cheap they'd rather steal our Star Generator than come up with one of their own. Well, they're not going to get away with it.

Airlock

I was right about the door—I'm obviously in an airlock. This can't be too difficult now, except that I can't seem to get this interior door open, either. Stupid Sariens—they can't even build a spaceship right. Even with all its faults, the *Arcada* could run circles around this reject. I *know* this is the door, if I could. . . . Wha? . . . "Hey, Bucko! Watch where you're going!" Dumb robot—it just about ran over me. Whoa—wait a minute! I'll just slip through this door before it closes. "Thanks, pal. Hope your insides short-circuit. . . ."

Trunk Room

Well, now. What is this room here? Nothing much, I'd say. I wonder what that trunk's for? I'll just take a look inside and . . . nothing. It's empty. Looks like a good place to dump this jetpack, though. I doubt I'll be needing it again, and it's just too heavy to carry around. Besides, a jetpack might arouse suspicion. Wouldn't do me any good anyway. The darn thing's almost out of fuel now.

Weeeell, I can see by the looks of this place the word *janitor* isn't part of the Sarien vocabulary. Look at this mess—candy bar wrappers all over the floor, dust in the corners, pieces of junk like this trunk lying around. . . . Uh-oh, what's that?

Someone's coming! What am I going to do?! Uh . . . uh . . . I'd better think fast! Uh . . . uh . . . the vent! I'll climb through the vent! But it's too high! What can I use? Oh, the trunk—of course! I'll stand on the trunk to reach the vent! I've got to hurry and push . . . the trunk . . . to the wall and . . . open . . . the vent. Oh, no. It won't budge! It's shut tight! I'll have to pry it open. Guess I'll use my. . . . OK, here goes. Oh, gee. I hear them coming—they're getting closer—and Sariens don't ask questions! It's open! OK, Wilco, JUMP!

Air Shaft

Ohhhhh! I don't believe it—just in time! They're taking the trunk away. Boy, I'm glad I didn't hide in that thing—but my jetpack! They'll discover it! Oh, who cares? It's from Kerona. I don't have to get so EXCITED over everything. They won't know it belongs to me. They'll just think there's a Keronian on board. But I'd better be veeerry careful.

Now to figure out how to get out of here. Let's see. This is a big air shaft. Should I go up or down? What the heck—I'll flip. Heads goes up; tails goes down. . . . Tails, darn it. I don't care what it says; I'm not going down. I want to go up!

OK, now what? Should I go up or right? Who knows? I'm tired of playing this guessing game. I'll just crawl to the end of the shaft and see what's there. Hopefully, it will be the Star Generator.

Gee, it's hard to see through this grate—and the darn thing won't budge again. Stupid cheap knife doesn't do any good—I bent it prying open that other vent. Now what? I guess I'll have to kick it open. Yeah, that's it. I'll give it my best Captain Kirk judo kick! Haaaaayak! Hah! I did it!

Now to see what unknown awaits me on the other side. I'm probably stepping into an ambush. They're probably waiting for me—all weapons pointed at me. That's OK; I can take it. If I must die, at least I'll die with dignity, sacrificing my life for the good of all mankind. At least I'll be known as a hero by whatever remnant of life remains in the universe once the Sariens. . . .

Laundry Room

Hey, this looks like a *laundry* room. There's no one here—no army of Sariens, no anybody. Ahem. Guess I should be thankful for small favors.

This looks like a pretty heavy-duty washing machine—but only one? I knew this place was filthy. It figures the Sariens would be, too. There must be a thousand crewmen aboard this tank, and with only *one* washing machine. . . . Eeeeeewwwww!

Well, I'm obviously not going to find the Star Generator in here. . . . Uh-oh. I hear someone coming! Oh, no. Not again. Suddenly I'm not feeling very heroic. I do believe I will pass on this opportunity to become a pile of ash, even if it does mean dying in the line of heroic duty. Now where am I going to hide? There's only one place in here I *can* hide— the washing machine! Oh, please, please, please don't find me! Shhhh! Quiet, Wilco, unless you want him to hear you.

Why's he coming over here? He must see me. . . . Oh, no. What's he doing? He's going to turn the washer ON! Oh, my gosh! At least he didn't see me. I guess . . . *(glub, glub, gurgle)*

. . . that's better . . . *(gurgle)* . . . than being shot . . . *(glub, glub)*. . . . Ow! Whoa! . . . Ouch!

Good gosh, that dimwitted thing spit me out! I'm probably the cleanest laundry it will ever. . . . Hey, these aren't the clothes I was wearing. Look at this gaudy outfit. Where's a mirror? Shucks, there's no mirror in here, but I'll bet I can see my reflection in the washing machine door. I'll just close it and. . . . Ahhhhh! A Sarien! . . . Oh, uh, ahem. It's not a Sarien. It's just me. Boy, I sure scared myself. I'll have to be careful about that. This is a Sarien *uniform.* Now I look like . . . one of them! Ugh!

Oh, well. I may look shoddy, but at least I'll be disguised. This is probably the single best thing that's happened to me all day. Now I can make my way through the *Deltaur* without arousing suspicion. This is great—what a coup! Now to find out what's outside the laundry room.

At the Two Elevators

Well, this place is certainly well guarded. Hmmm, two elevators. I'll just nonchalantly make my way to one of them and hopefully land on a floor with *no* guards. The main thing is not to call attention to myself.

OK, why isn't this elevator working? I guess it's broken. Oops. Guess I should have known that.

"Ahem. You, guard! I see this elevator is still broken, isn't it? Have the mechanic look at it immediately—and that's an order! . . . What do you mean you . . . *(gulp).* . . don't take orders from me? . . . Er. . . . Well, I didn't say the order came from *me.* . . . Do I look like I just rolled off the turnip truck or something? . . . The *turnip* truck, you dipstick. Haven't you ever heard that old say— Uh, never mind. Anyway, I didn't say it was *my* order. I was ordered to order you to have that elevator fixed. . . . Well, *something's* wrong with it, 'cause it ain't working, is it?

"It goes to the *what* room? Oh, well. . . . Yes, you're right. I should have. . . . I mean, I *did* know that. I've been running a temperature lately. Don't know why the boss has even had me working, seeing as how it's made me real delirious and all. And the doc said it was highly contagious. . . . Yeah, *reeeeal* contagious, so if I were you, I'd stay away from me! See how bad off I am? *(cough, cough)* I couldn't even remember where the broken elevator was located, but I assure you, there is a broken elevator aboard this vessel.

"Well, nice chatting with you. I'll just be heading back to my quarters on this other elevator here. At least I remembered that it works, ha-ha. I should probably stop by sick bay first. . . . I think I've gotten worse in the past five minutes. . . . Catch you later, pal."

Whew! Close call there, but I learned some very valuable information. I can see right now that it's going to be to my advantage to try to converse with some of these guards around here. I'll just have to be careful not to say too much— I definitely wouldn't want to give myself away! I can always ask which way to the little boy's room. That's a reasonable enough question—I hope.

Star Generator Room

Well, now. I wonder where this hallway leads. That room over there—I wonder what's inside. I'll just walk in there with my head held high like I'm on a mission. If I look official, like I'm supposed to be there, maybe no one will notice anything. OK, here goes.

Leapin' lizards! I just hit the jackpot! Give that man a cigar—it's the ever-elusive Star Generator! And it's being watched like a hawk—but only one hawk. Hmmm, I wonder if he could clue me in on anything I need to know. Looks pretty serious about his work. Maybe I should crack a couple of jokes. That should help break the ice—and make him think I'm just one of the guys.

"Hi, guy. How's it going in here? . . . I know you take your job very seriously, but I had some free time on my hands and thought I'd come by to see how you're doing. You could use the company, couldn't you? . . . Well, uh, I could use the company—so what's new?"

Great. He's as bad as the Keronians. "Uh. . . . Hey, did you hear the one about the guy at the bar? Well, there was this guy at the bar in . . . uh . . . it was in Ulence Flats. . . . Yeah, that's it. You know that old dive back on the planet Kerona? Well, this fellow walked in and heard the bartender just abusing some guy who was ordering drinks. . . . Well, every time the guy at the bar would order a beer, the bartender called him a jackass. 'Hey, Jackass, what do you want?' 'OK, Jackass, one beer coming up.' 'Here, Jackass. That'll be two buckazoids.' Well, the fellow who walked in and heard this was just dumbfounded. He couldn't understand why the bartender was being so rude. So he just walked up to the guy and asked him, 'Say, mister, why's the bartender being so rude? Are you going to just stand by and let him keep calling you names like that?' And the guy at the bar replied, 'Oh, it's OK. Hee-haw hee-haw hee-halways calls me that!' Ha-ha-ha-ha-ha! . . . You get it? That's pretty funny, huh?"

OK, so your sense of humor isn't exactly razor sharp. "Maybe you'd like to talk about. . . . Oh, uh . . . you don't want to talk period. OK. Well, then. How about a big fat kiss? Ha-ha-ha. . . . Aw, come on. Pucker up. Don't be such a stick in the mud, poker face. . . . Well, I'll just be making my way back out the door. I was just trying to be friendly, you know. Nothing wrong with being friendly, is there? Well, see you later." Jackass. I wouldn't kiss a toad like you for all the princesses in the universe.

Deltaur Lower Hallway

Good grief. Sariens are worse than I thought. What a bunch of losers. That cretin can't even get a good joke. . . . Oh, great. There's another one waiting by the elevator. Well, forget it. I'm not going out of my way anymore to be nice to these clods. I'll just say the minimum necessary to get by without drawing too much attention to myself.

"Hey."

Well, that was easy. What bores. I ought to thank my lucky stars I wasn't born a Sarien. Now, on with my real purpose for being here, part of which is not to converse with any more Sariens. Hey, what the! . . . What is THAT? Some kind of crazy robot, no doubt. What is it, another kind of guard or something? Just stay away from me, you clambering pile of circuits. . . . Y-y-y-yikes! Get away from me! . . . Whew! It doesn't know I'm not a Sarien. Boy, that's pretty pathetic. I could get away with murder in this uniform.

Now, I saw a catwalk running the length of that room where the Star Generator's rigged up. Maybe I could get a better idea of what to do from up there. I'm sure this is the right way—I didn't see any other elevators other than the one I took to get here.

☐ Chapter 5

Deltaur Upper Hallway and Armory

I was right! This hallway does lead to the catwalk. Now what to do. . . . Maybe I could jump that guard and knock him out! Nah, he's probably a lot stronger than I am. Besides, I'd be too scared. I have to think. Think, Wilco, think. . . . Hey, what's that room over there? It looks like some sort of weapons storage room. Hmmm, I believe my luck just made a turn for the best. Getting a weapon is going to be a piece of cake with this Sarien uniform. I'll just march right in there like I belong here—that bucket of bolts behind the counter will never know the difference.

Hmmm, those look like hand grenades on the counter. Maybe I'll get myself a couple of those. . . . "Excuse me? . . . My identification card? . . . Uh, yes, of course I have it with me." *What* identification card?!

"Uh, if I didn't leave it back in my quarters, that is, heh-heh. Well, let me take a look in my pockets here. . . ." Whew, I found it! "Oh, by golly, I guess I did bring that card with me after all! Forgetful of me, huh?"

I can see this thing has the personality of a drop cord. But while he's gone, I'll just grab these grenades. . . . Oops! Here he comes. Darn it—I only got one grenade. Hopefully, I won't need more than one!

"What? . . . Only one weapon? But what if this one breaks? Look. . . . Yeah, yeah, I know the rules. Nothing wrong with trying to bend them a little every now and then, is there? Lighten up; I was only kidding, for crying out loud. . . . Thanks for the pulseray. I'll use it in good health."

Yeah, and the first place I'd like to use it is on you, metalhead. This place is such a drag. Well, now. I see Mr. Personality the guard is still standing in the same position as when I left him—with his back turned to me. Should I shoot him? Noooo, I've got a better idea. I'll just take this little grenade here and. . . . All right! Good shot, Wilco! He's out like a light. . . . And hopefully, all of his lights are out for good!

Unfortunately, I used up my one and only grenade. You know, that dumb clunker in the armory doesn't know me from Adam except by my I.D. I bet I can get that other grenade if I walk in there again and ask for a weapon. What the heck—it's worth a try. I'll grab the other grenade as soon as he leaves the room. Hey, maybe he'll even give me a second pulseray.

"Ahem. I need a weapon. Sure, here's my I.D." Beat it, Bozo. I have a grenade to swipe! There he goes, and here I go. . . . Got it! Yes, sir. I'm in business now. I may not need this baby, but I'd rather be on the safe side just in case.

"Yeah, well, so what if I already have a pulseray. Don't you have any other kinds of weapons back there? What kind of cheapo armory is this anyway? I can have two weapons if I want, you rusted-out sack of wires. Can, too. . . . Can, too. We'll see about that. I'll go ask for permission. . . . Don't tell me when I can come back. I'll come in here anytime I please. . . . Hey, your father's married to a manual can open-er!" El jerko.

Finally, I'm actually getting somewhere. Now I just need to keep calm and act completely normal. No one should suspect that I'm anything other than a Sarien. I've got to get to the Star Generator room in a hurry, though, before someone else walks in and sees the guard lying face down on the floor! I will remain calm. . . . I will remain calm. . . . I will re—Whoooooooops! Oh, no! These blasted slippery floors! My helmet! Where's my helmet? I can't pass as a Sarien without that helmet! The minute one of those red-suited . . . or one of those clicking heaps of . . . I'm a goner!

The pulseray! At least I'm armed! I'm not taking any chances, though. The minute I turn any corner, pass through any door, or walk into any section of this ship, I'm firing away. This time I *will* be like a Sarien—shoot first and ask questions later. It's my only hope! I have to take it slowly, too. I'll just tiptoe my way to the Star Generator room.

```
Score:187 of 202                    Sound:off




     This small item is a
     Pulseray. It delivers
     concentrated pulses of energy
     in the direction it is
     pointed.




>_
         F6 to fire Pulseray
```

Don't wait to see the whites (or reds or greens) of their eyes. Shoot the air if necessary; it's the only way you'll survive.

A robot! All right. Well, take that, you nincompoop! Uh-oh. The pulseray doesn't work on robots! Y-y-y-yikes! I'm getting out of heeeeere! . . . Whew! It didn't follow me in here. Thank goodness. I hope it's gone now. I *have* to get to that Star Generator room on the double! OK, here goes again. This time, I'll march in as though I'm facing a front line of fire!

"Take *that* and *that* and *that* and *that* and *that*, you jerk!" OK, one Sarien down, but how many more to go?

"Take *that* and *that* and *that* and *that* and *that!*" Another Sarien down, and I'm almost there! I'd better be prepared to fire in case some of them are waiting in the Star Generator room.

Star Generator Room

"Take *that* and *that* and *that* and *that* and *that!*" Thank heavens! No one's in here. Now I have to figure out what to do with the Star Generator. I'll just give it a close— OUCH! Wow-wee-zing! That thing has a force field around it! I forgot about that! Now what do I do? There has to be a way to turn it off.

Wait a minute—I remember seeing some sort of contraption on this guard when I was in here earlier. I'll bet that's the control to the force field. OK, you stinking, nasty Sarien, I'm not going to enjoy searching you. Yuck! Just don't give me cooties. . . . Here it is! This is it! This must control the force field!

Boy, woooo, these Sariens are real geniuses. Gee, I don't know if I have the brains to figure out how to operate such a complicated piece of equipment! Let's see. Should I press ON or OFF? Oh, heck. I'll just throw my luck to chance and press OFF. Well, bless my soul. I do believe that force field just turned off!

OK, Wilco, enough with the clowning around—this is serious stuff. Now that the force field is off, I can take a closer look at the Generator. Let's see. . . . I have to enter a code. A code . . . a code. . . . Oh! What was it that cartridge said? It's been so long. . . . I don't know if I can remember! Think, Wilco, think! It was four numbers! I remember. . . . *Sanitarian.* . . . *Equality.* . . . *For-Ever!* That's it! Now, what were those numbers? . . . Sanitarian is . . . six. . . . Equality is *eight.* . . . For—that one was tricky—is *five.* . . . And Ever— that one's easy—is *eight.*

The code numbers are *six, eight, five, eight!* I remember! Now to enter that code, and—voilà! The Star Generator is now set to self-destruct—and I've got to get the heck out of here before I self-destruct with it!

I'll bet dollars to doughnuts there's a Sarien waiting for me outside that doorway . . . but here goes anyway!

"Take *that* and *that* and *that* and *that* and *that!*" I knew it! There was one—*was* being the operative word here. Now into the elevator and up to that floor with the entrance to the not-so-broken elevator. OK, I have to be prepared to fire again the moment I step foot out of this elevator! I don't even have time to be afraid—that Generator's going to go up like a fireball any minute now.

At the Two Elevators

"Take *that* and *that* and *that* and *that* and *that!*" Oh, wo-wo-wo-wo—yikes! It's that walking neon sign again, and it doesn't respond to pulserays! I'll make a dash for the elevator anyway. . . .

I did it! Funny how I'm willing to take a few more chances here at the end. I guess I'm really into this hero thing now—as if I had any choice!

OK—down, down, down she goes. Where the elevator stops, nobody knows! . . .

Escape Room

This is IT! This is the escape room, and my ticket out of here is parked right over there. I'll just make my w-w-waaaay. . . . Whoops! Whoa! I'd better be careful here! I didn't realize this escape ship was suspended over some bottomless pit! Whew! How awful if I had come this far only to plunge to my death just as I was about to escape—and after everything I've been through! I'll be extra careful about getting into this ship, but I have no time to waste.

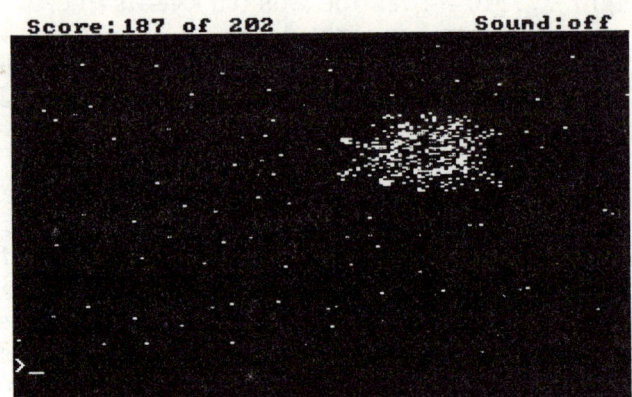

*Mission accom-
plished when you
see this sight.*

I'm in! Yes! Now let's get this baby out of here! Let's see. I'm still not an engineer—guess I never will be one. My guess is to push the button that says *Power.* That's what I did in the escape pod. OK, here goes.

I did it! I made it! Yahoooooooooo! I'm out of here—and out of danger! And whoa, look at the *Deltaur* now. . . . It's nothing but a thousand points of light—that are about to go out *permanently!* I can't believe it! I did it! I'm a hero! Let's hear it for me! Yaaaay! I just can't WAIT to go home now. . . .

The Planet Xenon

"Hey, Fly-Boy. Can you believe this? We saved the universe! You and I did it together—although I did most of it! Look at all these people cheering us on! They'll never look at me the same way again. They'll never think of me as *just* a janitor again! And look, there go some of my fellow legends! Gee, I wonder why they didn't help out? They're real good at this stuff. Oh, well. . . . Check you later, guys—live long and prosper!

"Oh, look, Fly-Boy. They're bringing me a trophy! Can you believe it? They're honoring me, Roger Wilco, with a . . . a . . . gold mop? Well, uh, gee, I guess that's not so bad. Ha-ha—it's probably just a joke—or a reminder of what I *used* to be . . . but not anymore. Yes, that's OK. I'll be a good sport.

Game Points Earned

Action	Points
Climbing through the vent	2
Conversing with a Sarien	1
Entering the self-destruct code	10
Escaping the *Deltaur*	3
Finding the trunk	1
Getting one gas grenade	1
Getting the pulseray	3
Getting the remote control	3
Getting the second grenade	1
Hiding in the trunk	3
Hiding in the washing machine	5
Kicking the grate in the air shaft	1
Killing the guard at the Star Generator	5
Kissing a Sarien	1
Making it to the escape room	1
Saying yes to *King's Quest II*	5
Shooting one Sarien guard	3
Turning off the force field	3

Object Locations

Object	Location
Escape ship	Escape room
Identification card	Sarien uniform
Pulseray	Armory
Remote control	On guard in Star Generator room
Two gas grenades	Armory

Chapter 6
Vohaul's Revenge:
Close Encounters of the
Sludge Kind

Poor old Roger Wilco! And you thought your days as a janitor were over. So the Xenonites forgot—uh, rather quickly— that you saved the entire universe. Someone else, unfortunately, remembered. He's the one to look out for. Who cares about the boss anyway?

Stop being such a pushover. If you had any gumption you'd have told those clowns at the Sanitation Office to go jump after your last adventure. Sanitarian equality—that's a laugh—and the joke's on you, according to those unappreciative higher-ups.

For people like them, getting a job done is the only thing that counts, but you should know there's more to life than that. At least one would *think* you would know, considering how recently and how often you almost bought the farm! Those kinds of experiences do have a way of putting things in perspective for most people—you know, reevaluating priorities and all.

Well, on second thought, you're *still* no James Bond. Maybe you *are* one of those people whose greatest foresight is what you're planning to do on the next coffee break. If that's the case, bud, you're in for some hard times. This adventure ain't over 'til it's over. Someone's out for revenge— on you!

Total Possible Game Points: **250**

Xenon Orbital Station 4

Mission: Clean out the shuttle on Orbital Station 4.
Total Points: 10

Orbital Station 4

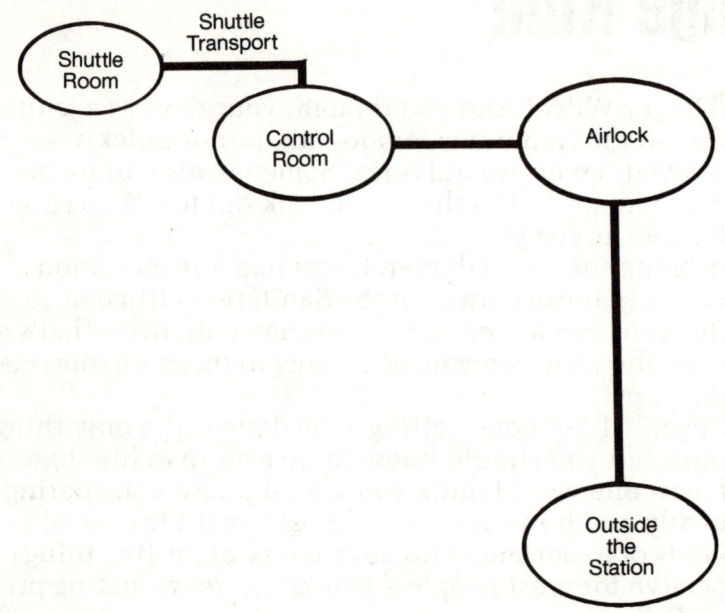

Inventory

Cubix Rube
jockstrap
order form
translator

Outside Orbital Station 4

Sweep, sweep, sweep—day in and day out—that's all I do.
Tote that barge, lift that bail. Some appreciation. I saved this
blooming universe from the hands of the evil Sariens, but
can't get any respect. Nooooo. They stick a gold mop in my
hand—rolled-gold plated at that—and tell me to get to it the
first day after my return. From a ticker-tape parade to mop-
ping floors again.

"Oh, Roger, we have a new assignment for you. You're
really going to like the Xenon Orbital Station 4. And your
new boss is well aware of your impressive history." Blah,
blah, blah. Oh, he was well aware all right. The guy's been
after me from day one. "So you think you're some kind of a
hero, huh? Well, let's see how heroically you can sweep the
trash out of that air shoot—and on the double! And don't get
any high and mighty ideas about how you're too good to do
this work, Wilco. You may have been a hero yesterday, but
you're a janitor today."

Once a janitor, al-
ways a janitor. So
your boss forgot
you saved the
world from the evil
Sariens. Unfortu-
nately, someone
else remembered.

Nag, nag, nag. That's all they do, day and night. I never
get a moment's rest—except in the hall closet. And this job
doesn't pay diddly-squat. Well, maybe I'll show them.

Not again. Nag, nag, nag. Here we go again. What does he
want this time? I hate this stupid watch. Some present. It's
just another way for the boss to nag me even when he isn't
around. . . . Hey, my broom! Darn it. They're not taking it out
of my salary this time. It's not my fault the beeper startles me
every time it goes off.

"What is it this time, your high— I mean, boss? Aw, get off my back. I wasn't getting smart with you—it was just a joke. Yeah, yeah . . . I know where the shuttle is. I've only been on that thing a million times. . . . Well, I'm sick of cleaning out that junk heap. Why don't they clean up their own mess? All right, all right, I'm going. You don't have to yell."

A nice day to you, too, Bucko. That jerk's always been jealous of my fame—I know that's why he's on my case all the time. You would think I'd have gotten a little reward money for risking my life to save this lousy universe. No reward money, no stereo set, no new CD player. Heck, I can't even get a date. I thought heroes were supposed to be desirable. Ha, that's a laugh—not when they're rewarded with a mop. Time to go in now. Woe is me. "Life is bare—gloom and misery everywhere. . . ."

Airlock

Well, time to change clothes for the umpteenth time today. "Clean the outside. Clean the inside. Clean them both." I tell you, I CAN'T TAKE IT ANYMORE! This job is the. . . . Heyyyyy, someone's been in my locker. I can tell by the way it's closed. I'm really getting sick of this. Where's my key? I know I left it in a pocket here. They'd better not have taken my picture of. . . . Oh, here's my order form for the Labion Terror Beast whistle. I keep forgetting to mail it. Gee, I've always wanted one of these. I'd better hang on to this. I'll mail it first thing after cleaning out the shuttle.

Heyyyyy, not only has someone been in my locker, but someone's been wearing my uniform. This gadget doesn't belong to me. What the heck do I need a translator for? That does it. I've had it with this. I'm telling the boss. They'd better stop messing with my stuff. There's no telling what's been taken from my locker.

Well, everything looks like it's here, except for this stuff—an athletic supporter and a Cubix Rube. Wait a minute—this stuff belongs to that nitwit on the night shift. I told them to give us different keys for these lockers. That birdbrain never remembers which one is his. Wait until I report him. I'm taking this stuff straight to the boss.

Control Room

Boy, I despise walking through this control room. Everyone in here acts like they're better than I am. Oh, no—not that jerk again.

"Don't nag me now; I'll get to it when I'm ready. Why don't you worry about your own job for a change! . . . I don't care what you tell the boss. I'll just tell him about you." Creep.

Oh, there's Gladys, the communications officer. Wow, she's really swell. I've just never been able to get up enough nerve to talk to her. . . . I wonder if she'd go with me to the Monolith Burger. I've got to make myself ask her, but the girl acts like I'm not even alive. Maybe that's just a game. Maybe ol' Gladys is playing hard to get.

Well, I made myself do all sorts of things when I was faced with sure death by the Sariens, so surely I can make myself do this. . . . Hey, that's it! I'll remind her of who I am! I'll bet she doesn't even realize that I'm Roger Wilco, the one who saved Xenon and every other planet in this galaxy! Oh, well. It's now or never!

"Hi there, Gladys? I don't believe we've actually met before, but I've seen you around a lot. I'm Roger Wilco. You, uh, may have heard of me. . . . Well, don't be shy now. . . . Uh, well, I guess you are shy. OK, uh, I was wondering if you would be interested in going out to dinner with me. I thought we could go over to the Monolith Burger, and . . . uh . . . Gladys? Ahem. Well, I'll definitely take that as a no. OK, well, see you later. Maybe some other time, huh? I'll check back with you when you're not so busy. . . . I have to get back to work now. See you later. . . ."

Holy cow, I just made a fool of myself. That girl didn't even look up from her post! My face must be a thousand shades of red by now. Oh, I'm so embarrassed! No wonder she's never looked at me before—her nose is cemented to that communications equipment. Wilco, you really blew it this time. And she probably *hates* fast food. Those intellectual types are like that. Maybe I should have tried a different approach. I know, next time I'll *call* her—she's into communications and all. . . .

I just need to get to that shuttle as fast as possible now before everyone starts looking at me. Boy, I feel like a real dunce. I'll bet nobody in this entire room eats at the Monolith. Oh, well. I've got a job to do—at least it will help me take my mind off Gladys for a while.

I sure wish they'd come up with another form of transportation to the shuttle bay. I always feel like I'm being shot out of a straw like a spit wad or something when I ride this thing.

Shuttle Bay

OK, now for the dirty work. Gee whiz, I sure hate cleaning up after those morons. I thought this was supposed to be a decent post for a janitor. Nothing's ever. . . . Hey! What the. . . . Oh! Ouch! Oooo! Ow! Ohhhhh. . . .

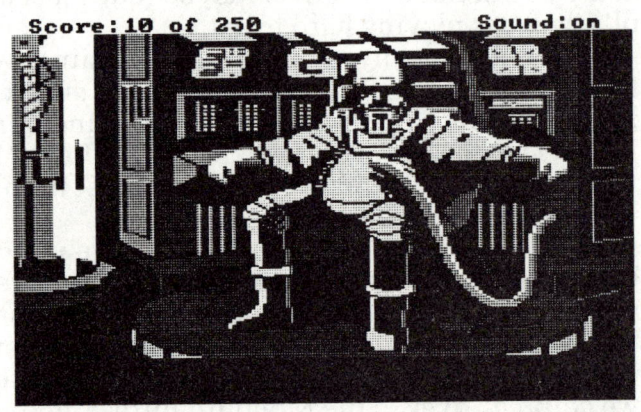

He would have a name like Sludge. Look at how he sits. Ugh!

Huh? Wh-where am I? Oh, my aching head! Ooooo, someone threw a good one at my left temple. . . . What in the world?

"Who the heck are you, or rather *what* are you? Sludge what? Mister, you look like sludge. Where's the door? My head hurts, and I've got a shuttle to clean.

"You're what?" Uh-oh. I think I'm in trouble. This character's got a score to settle with me—and I thought I made the universe safe. . . . Geez, safe for everyone but me.

"Yeah, I've heard of Labion. I even plan to order. . . . Hey! You don't have to yell! I can hear, and you're killing my ears! I've got a headache, not that it means anything to you.

"You're planning *what?!* An infestation . . . of . . . oh, no! You wouldn't! You can't mean that! Do you know what you're saying?! INSURANCE SALESMEN?!!! On Xenon?!"

Oh . . . my . . . gosh. It *is* a clone. Look at that . . . that . . . *thing!* This can't be. What kind of hideous, twisted mind would conceive of such a plan? "You really are *sick,* you overgrown tub of lard. And you make *me* sick! Sludge is right, you dirt ball. You're sewage . . . garbage. . . . You're the *scum of the earth!* I hope you get yours, Vohaul, and I hope you get it good. And I may just be a janitor, but I can be a whole lot more, too! If I have anything to say about it, you'll rot in. . . . Hey, where are you taking me?! I don't want to leave the shuttle. I don't want to! Let go of me. . . . Tell me the truth, you walking heap of refuse; did my boss have anything to do with . . . your . . . finding . . . me? . . ."

Game Points Earned

Action	Points
Changing clothes	1
Entering the airlock	1
Entering the shuttle to clean it	5
Getting the Cubix Rube	1
Getting the jockstrap	1
Seeing the boss on your wristwatch	1

Object Locations

Object	Location
Cubix Rube	Locker
Jockstrap	Locker
Translator	Clothes pocket
Whistle order form	Clothes pocket

The Planet Labion

Mission: Survive the obstacles you're about to encounter.
Find a way to get off the planet.
Total Points: 156

The Planet Labion

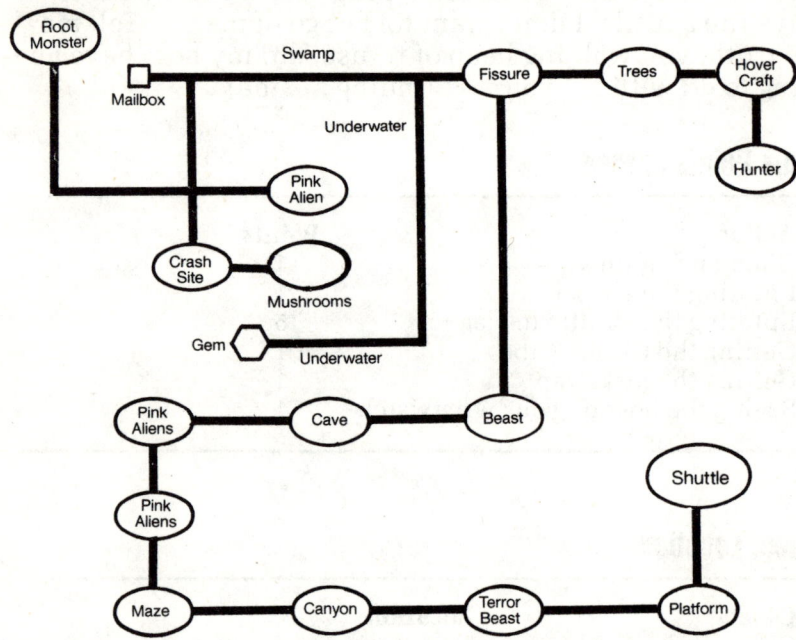

Underground Maze

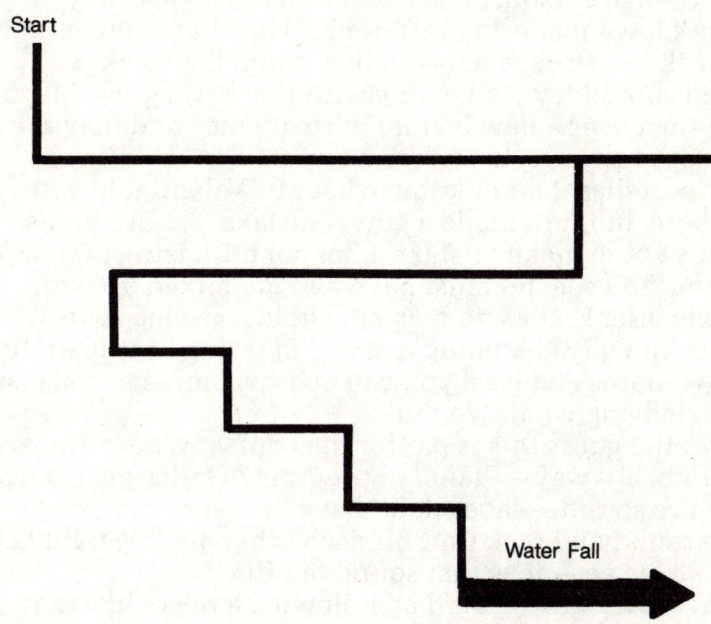

Start

Water Fall

Inventory

berries
cage key
gemstone
keycard
rock
rope
spores
whistle

On the Hovercraft

Gosh, this isn't very funny. Maybe I should have been nicer to that Sludge person, or whatever he is. No—no way. It wouldn't have made any difference. He'd have thought I was weaker than I already am—well, not that I'm weak really. Maybe I should try conversing with these two guys. Maybe they'll turn over a new leaf if I'm really nice and they see I'm no threat to anyone. It's worth a try!

"Hey, fellas, I don't know what Mr. Vohaul told you back there, but he's made a grave mistake. . . . But that's OK—it's OK to make mistakes! I'm not criticizing him or anything. You see, he must have had me mixed up with someone else. He has no reason to hold a grudge against. . . . Hey, are you guys listening to me? I'm trying to explain the situation here. The least you can do is pay me the courtesy of acknowledging what I've said.

"Well, I guess this is par for the course. Where are you guys from, anyway—Planet of the Apes? Ha-ha-ha. No wonder you're so rude—apes don't have manners. . . . You just stand around and pick flies off each other all day, right? Oo oo, ah ah, ee ee—does that sound familiar?

"Yeah, it figures you'd be following a reject like that Slime, Slop, Sludge, *whatever* his name is. Monkey see, monkey do, right? Ooooh, struck a nerve there, did I? So what are you going to do about it, bang on your chests for a while? . . . Oh, gee. I'm really scared! . . . Awwww, now what? Did you guys forget to fill 'er up or something? You two must have been up to some monkeyshines when you were supposed to be taking care of this craft. Well, Mr. Slime, uh Sludge, isn't going to be too happy about that, is he? Doesn't make any difference to me. I'm a goner anywaaaaaaay. . . ."

Crash Site in the Labion Forest

Great balls of fire! I don't know how many more poundings my head can take. . . . Huh? I don't believe it. Looks like I'm the only one left of this trio. Those gorillas sure made my fall a heck of a lot easier—talk about a twist of fate. And what's that beeping noise? It's sure getting on my nerves—and it's coming from the hovercraft. Oh, no! What if it's a homing device? I'd better shut it off before another dynamic duo decides to hover my way.

Now where should I go from here? I know I'm on the planet Labion—I wonder if there really *is* such a thing as a Labion Terror Beast. Hmmm, I bet I'm going to find out. Other than that, I really don't know anything about this planet. One thing's for sure; I don't plan to hang around this crash sight. Before I leave, however, I need to give these compatriots a once-over—see if they're carrying anything useful. I could sure use one of those weapons, but it looks like they've both been destroyed.

OK, you first, big guy. What's this? A monkey wrench. . . . It figures. I don't need that piece of junk. Nothing else on him. OK, King Kong, you're next. . . . Hey, a keycard. These things are all over the place—I'd like to be the guy who invented them—he must be a millionaire by now. Well, that's that. At least I'm still alive, but I'm still in danger as long as I'm on this planet.

Huh! Wh-wh-what was that? Someone or something was in those bushes over there! . . . This place is starting to give me the heebie-jeebies! Maybe it's just my imagination, but I could swear I saw little eyes glowing at me through those bushes. I feel like I'm being *watched*. I'd better get out of here—but which way? The forest is so dense!

I'll have to make a decision. I'll go east. . . . Huh! What's that dark green patch over there? It looks very strange to me. Maybe I should get a closer look. . . . On second thought, I believe I'll pass on it. OK, here I go. I just hope there's nothing sneaking up behind me in this creepy place. . . . I hope there are no Orats or apes or droid spiders—oh, my! Orats and apes and droid spiders—oh, my! ORATS and APES and DROID SPIDERS—OH. . . . Hey, what's that? A giant mushroom? I wonder if it's edible—I could stand to get the subject of food off my mind for once.

I'll just take a closer look. . . . Or maybe I won't. What was that I read in my last issue of *Janitor Today*? Let's see; it was something about a very acidic type of mushroom that's being used in some of the new wall cleaners. I don't remember everything it said, but it seems like the mushrooms are supposed to be big—like these. Well, Wilco, on second thought, let's not examine the mushrooms and say we did. They may not be the same strain, but then again, maybe they are—and they could be deadly!

Great. It looks like I'll have to leave the way I came in—there's no other way out of this clearing. There's that dark green patch again—I'm still staying away from it. Something about it just doesn't look right.

Now where should I go? Actually, it looks like I may not have a choice. The forest is so thick, and it looks like there's only one way out of this area, too. Fine, I'll just be on my way. Thank goodness no one's discovered the wreck yet—they'll be looking for *me*!

Forest Clearing with Large Trees

Now where should I go? I see several paths leading out of this area. . . . Yikes, what was that! It sounded like a scream! Oh, that isn't very g-good. It sounded like it was coming from over th-there. I'm so scared—I'm frozen in my tracks! What should I do?

OK, Wilco, you've got to get hold of yourself. I can do this. I've done it before—I just hoped I would never have to do it again! Suddenly my job back at Orbital Station 4 sounds like a million buckazoids. What I would give to be back there, minding my chores, safe and secure, scrubbing away on someone's floor, away from all danger. . . . Oh, well. That's not going to happen. I'm here, and if I want to survive, I at least have to muster up enough courage to investigate that scream. . . . Ahh! There it is again!

S-someone's in an awful lot of p-p-pain. It's coming from over there! If I go very quietly, maybe I won't be noticed. It's probably one of Vohaul's cronies torturing some poor, innocent creature. OK, I'm almost there. . . . I must be quiet—no screaming allowed!

I don't see any orangutans swinging from trees. . . . But there's *something* swinging from the tree! What is it? It's pink! It's a little pink alien! Actually, it's probably a little pink native—I'm the alien here. I've never seen one of those before. I wonder what he's doing in that tree.

"Hi there, little fellow. Can I help you?! Are you stuck? Oh, I guess you don't speak English. Well, I have a translator here. I could probably understand you if you wanted to say something! . . . OK, well, you don't have to say anything. It looks like someone laid a trap for you.

"I'll untie you. I'll be glad to help! Hold your horses; I'm almost done. There you go, little fellow. . . . Hey! Wait! Come back! I won't hurt you; I promise. I'm not the one who set the trap! I wouldn't do that! Well, bye now, if you're in such a rush!" Oh, well. I guess he was afraid of me. At least he wasn't out to *get* me. I could certainly have used a friendly face for a change.

That's a pretty thick shrub he climbed through. No way I could get through that mess. I think I'd better stick to my original path. . . . I know I saw something in the distance out there. It looked like a mailbox, of all things.

OK, I need to check this out. What would a mailbox be doing out in the middle of the forest? If it really is a mailbox, I can send off for that Labion Terror Beast whistle I've been wanting to order. The only problem, of course, is what to use as my forwarding. . . . Uh-oh! What's that noise?! It sounds like one of those hovercraft! I can't let it see me! They'll kill me! But where can I hide? Y-y-yiiiikes! Oh, please don't let the hovercraft see me!

Whew, that was close, thanks to this tree! Those things must be all over the place. I have a feeling they've already discovered the wreckage—and now they're looking for me! I must be careful. Now, where was I? Oh, back to the mailbox. I don't know why I'm even bothering with this—I probably won't be around another six weeks or however long it takes for the whistle to come.

Forest Clearing at the Mailbox

I was right—this is a mailbox and an express one at that. It's got to be faster than the one on Orbital Station 4; that's all I have to say about it. OK, I'll just insert the envelope in the slot and be on my. . . . Whoa! What in the world? . . . Something just popped out into that tray. No, it can't be. There's no such thing as mail that fast—and never will be in my lifetime. I'll just see what it is, though. . . . I don't believe it! It's a Labion Terror Beast whistle! But how? Boy, talk about service! Whoever runs that service is sitting on a gold mine!

A real Labion Terror Beast whistle! I've *always* wanted one of these, and now it's mine—all mine! I'll just put it in my pocket for safekeeping. Who knows when I may need to call a Labion Terror Beast for help—since I'm on the planet Labion and all. . . . Ha! Fat chance! There's no such thing!

Now where should I go? I see a swamp in the distance there, but I'm not too keen on getting wet. I'll just hop down this little embankment here and see if there's another path I can take.

What are those strange-looking things on the ground—some type of spore maybe? I'm a janitor, not a botanist, but they sure look like spores to me. Saaaay, I believe I remember reading about these things somewhere—if those are what I'm thinking about. Yeah, that's it—one of the custodians at Orbital Station 4 brought some back from his vacation. He left one lying on the floor in one of the closets. When his boss went to look for him, he accidentally stepped on the spore, and it paralyzed him! Yeah, that story was in the paper the other week! I remember because I was wishing I had one of those spores myself! And these look exactly like the ones pictured in the newspaper.

Boy, am I glad I read that story. Now these things could really come in handy, but I'd better be careful. I don't want to get paralyzed myself! I'll just take one and veeery carefully place it in my pocket. There! . . . Whoops! I almost lost my balance! I sure as heck don't want to step on one of those things! I'll just make my way clear of them. . . .

Root Monster

Let's see; this path leads to somewhere. I hope it's a path around that swamp. I really have no desire to get all wet and mess up my hair with no blow dryer. . . . Geeeeeee! That's the oddest looking tree I've ever seen in my life! Labion sure has some strange-looking life forms.

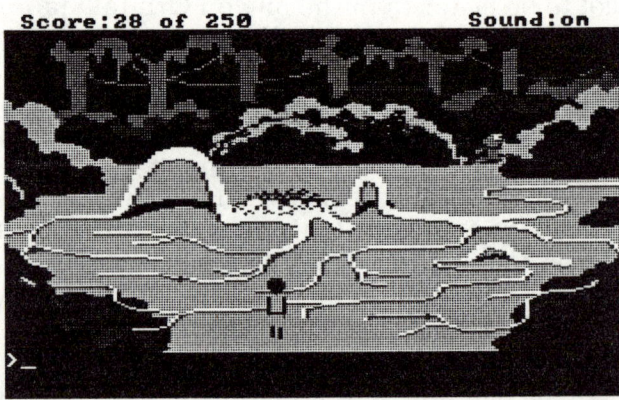

Step too close to this "tree," and you'll find yourself uprooted. Those berries in the background would make a nice waterproof cologne.

It looks kind of gross, to tell the truth. I think it's moving! Hey, there's the pink alien picking berries off those bushes. He's rubbing them all over himself! Well, that's an odd way to eat berries. They must smell really good for him to be doing that! "Hey there! Yo! Hey, it's me, Roger Wilco! I untied you, remember? Hey, where are you going again? Don't run away. . . ."

Shoot. That little guy's really afraid of me. Oh, well. . . . Saaaay, I wonder what kind of berries *are* on those bushes. Maybe some very good ones like strawberries or blackberries or boysenberries or any kind of berries. And I mean *any* kind would taste good right now! I don't know about walking over this . . . this tree thing, though. Something about it just isn't right. I'll definitely have to tiptoe through this maze—very carefully and very slowly.

OK. . . . I'm almost there. . . . Almost there. . . . Whooooops! Whew! I lost my balance there for a minute. Don't want that to happen—I just have a feeling that there's more to this tree trunk, with all its roots, than meets the eye. It's almost as if it's waiting for me to touch it so it can reach out and grab me with one of its long, tentacled roots. Oooh, I don't even want to think about it! I'm starving, and I have to get to those berries! . . . Just one more little itty-bitty turn and. . . . Made it! Great! I just hope there's an exit path through these bushes.

Well, I don't know what kind of berries these are, but they look awfully good to me! OK, Wilco, pick berries! I'll get as many as I can and stuff them in my empty pockets. . . .

That's about all I can hold for now. I can always come back for more if there's an easy way out of here through these bushes. . . . Gee, I don't see any openings in the shrubs. How do those little guys do it? Oh, no. Don't tell me I have to make my way back through that root thing again. Well, that figures. I guess it was worth it to get these berries. I'm soooo hungry.

Before I go, I might as well eat. I'll just stuff a handful of these juicy yum-yums in my mouth and. . . . Ugh! Barf! Whooo! That smell! I can't even get these things close to my face. They're so pungent that my eyes are stinging! I can't eat these berries—they smell horrible! Imagine how they must taste!

No wonder that little pink guy wasn't eating them. But why was he rubbing them all over himself? Hmmm, I wonder if he knows something I don't. There must be some significance to that action. Well, I'll just do the same thing—as soon as I get back through this root maze. What a disappointment. Oh, well. "OK, roots, here I come! . . . ."

Well, that was much easier than the first time through. Maybe the root monster smelled the stench emanating from my pockets and decided to let me go through. Heyyyy, maybe that's what these berries are for—warding off hobgoblins! That gives me all the motivation I need to rub them all over me, just like the pink guy did. I'll have to hold my breath to do this. . . . Uggggh. The smell is really the pits. These berries smell worse than ol' Sludge himself. I'll bet he eats them for dinner every night.

OK, I'm covered in berry stench now. Hope it works like a charm. All I know is that I haven't found a way to bypass that swamp, so it looks like Roger Wilco is going for a swim. I sure hope the berries are waterproof.

Forest Swamp

This place is really bleak—it's spooky. I think I'm alone, and yet I still feel like I'm being watched. I don't like this feeling at all. I'd better make haste getting through that swamp. No wonder Labions use hovercraft—there's barely any place to walk on this dismal planet.

There's no telling what lives in this swamp beneath the surface. I'd better not think about it. I just wish I could swim faster through this shallow part. Huh! What was that?! I heard something move. . . . I don't see anything. . . . I swear I think I'm being watched. Could it be just my imagina—WHOOOOOOOOOA! AHHHHHHHHHHH! . . . Ack! . . . (glub, glub, gurgle) Acccck! (cough, cough) Wh-what . . . (cough) . . . happened?! (cough, cough)

(cough, cough, cough) It tried to eat me! A monster in the water tried to eat me! I felt its teeth and its hot, slimy tongue on my head! Accck! It must be the berries—they saved me! That thing was about to have me for its first course, but it must hate the taste of those berries! Oh, thank you, little pink fellow, wherever you are!

Boy! That was close! I'm so glad I rubbed those berries on me—I'd be in that creature's intestines. . . .

Eeeewwwwwww! I can't think about that right now. I'll think about that tomorrow. Right now I have to concentrate on getting out of this cesspool. I feel like I'm walking on Vohaul's stomach! Ugh!

Whoops! I've come over a deep part here. There's something strange about this. I definitely feel an undercurrent here, but it's a vertical pull—as though there were a hole in the swamp. This is very strange. I wonder if there's a cavern of some sort below. I may not be James Bond, but I *am* a good swimmer. What can it possibly hurt for me to take a little dive underneath and see if it leads anywhere?

Beneath the Swamp

I'll go for it. I'll just take a deep breath . . . and dive! . . . Wow, I can't believe how easy it is to see down here. Hmmm, it's a cavern all right—of some sort. The walls are very smooth and slippery. I wonder how deep it runs. I'll swim as far as I can. . . . Oh! I've hit the bottom—but there's an opening running west. I still have plenty of air. I'll keep swimming until. . . . Ouch! This opening is just a bit narrow! . . . Hey, I see a light. There must be an opening up ahead.

Whew! I made it. I was starting to get a little lightheaded there for a moment. Wow, what's that?! Some kind of gemstone. It's beautiful. It's so faceted it actually emits light. I'll bet this gem's been in here for centuries—but how in the world did it get here? I'll bet some brave soul put it here for safekeeping, probably to be given as a gift to his bride in waiting somewhere—only he never returned. What a sad story. I'll bet this thing's worth a fortune. You'd better believe *I'm* not going to leave it down here. This little jewel is going in my pocket!

Now to get out of here—there are obviously no other openings through which to escape. I'll have to go back the way I came in. Time to take a deep breath again . . . and dive!

Whew! I made it again. Thank goodness for those swimming classes at the Y. Now to get the heck out of this sewer— who would have thought something so beautiful would be hiding beneath these murky waters?

Forest Beyond the Swamp

Finally, I'm out of that dreary swamp. I hope I never have to go through that again. Let's see what's up ahead. . . . Hmmm, a fissure in the ground. This planet just flat isn't made for ground travel. How the heck am I supposed to get across that thing safely? It looks very deep. Maybe I can pull a vine from one of these trees and swing across it. No, wait a minute. This old hollowed-out tree trunk might do the trick! I could climb to the top of it and jump. The fissure is a little narrower here—I'm sure I can make it!

A rope, Boy Wonder! That's the only way we'll find out what lies beneath this hollowed ground!

    I'll just put my foot on this stump and then another foot on the next stump and then. . . . Ahhhhhh! Oh, my gosh. I just about had a heart attack! This lousy, falling-apart. . . . Hey, wait a minute—I can get across now! This old rotted trunk just made a bridge for me to cross. I should have thought of that to start with.

    Saaaaay, now that I'm over this hole in the ground, I can see a ledge and a cave opening down there. I wonder where that leads. Hmmm, if only I had a rope to tie around this log. Then I could climb down and check out that cave. Oh, well. Too bad I don't have a rope. Maybe I can tug a vine loose from one of those trees over there. Of course, a knife would come in very handy there, too, but maybe one of the vines will be loose enough to pull off.

    OK, I see lots of vines hanging from these trees. Now, which one. . . . Whooooooa! Hey, what's happening? Oh. . . . I see stars. . . .

The Ogre Hunter's Camp

Oh, my aching head. This planet has brought me nothing but grief so far—except for my whistle and gemstone—and another splitting headache. Ouch, Ooooo, where am I anyway? Please tell me I'm not back on that shuttle with the Sludge. . . . No, I can see I'm not. . . . I'm in a . . . CAGE?

What's going on here? Huh? Who's that humpbacked Goliath slouched over the campfire? He's an ugly piece of work. Those vines—he must have set them as traps. That's how the little pink alien ended up hanging from his toes. Too bad he isn't here to rescue me.

I wonder what this ogre has planned for me. I'll bet I can guess. Well, I've got news for you, pal; I ain't hanging around for dinner—not as your guest and certainly not as your meal! Now if I can figure out what to do.

Put on your thinking cap, Wilco. This is another one of those life-or-death deals. . . . Hmmm. . . . I need to knock his block off somehow. But with what? There's nothing in this cage, not even a rock to throw. Maybe I have something here in my pockets. . . . I've got it! Who says I only have the foresight to make plans for the next coffee break? I read that newspaper article, didn't I? OK, you big buffoon, get ready to eat dirt!

"Ahem. Excuse me. Could you come over here for a minute please?"

Oh, great. Either he's deaf or he's just ignoring me. Well, I will *not* be ignored!

"Hey, you. Excuse me. I'm talking to you, mushbrain. Knock, knock. Is anybody there? What's the matter—don't you understand English? By the size of your head, I'd say they forgot to put in a brain. Oh, gee. If you only had a brain—nah, no thoughts will ever hatch from that head, no matter how many times you scratch it. And you'd still be here for nothing, your head all full of stuffing."

Oh, so he moves. I thought that would get his attention. Boy, look at that face. Lots of lights on, but nobody's home. This should be easy. I'd better hurry, though—he doesn't look too happy. That's it—come over here to the cage. I need you to be close enough so I can get your key.

"Get a whiff of this, Fido. . . ." Ha! It worked! And I can just reach that key hanging out of his pocket. . . . Oooo, coo-

ties—should have known. I'd better be careful. Now to un-
lock the cage.

Great, I'm out! I can't believe how resourceful I've become. No more janitorial slumming for me! And I'll just take
that rope while I'm here. I may need it later for that fissure—
or who knows what else. "Ciao, gizzard breath. May all of
your meals get up and leave."

You shouldn't rush in where angels fear to tread. Best to lay low and go west.

Beyond the Hunter's Camp

Now to find my way out of this gutter. Let's see. . . . Hey,
there's a landing pad in the distance. I see the Sludgettes taking off. I wonder if I could make it over there. But that cliff—
I don't know. I hesitate to even walk over there. Something
about this picture just doesn't look right. I might be too visible if I try to climb down that cliff. If I can see the landing
pad from here, they would see me for sure. And those hovercraft are much faster than my feet—not to mention their
weapons.

OK, I'll ditch that idea. I'll just lay low and make my way
through those thickets. There's bound to be a way I can
sneak through the forest to get to that landing pad. One way
or another, I'm getting off this planet!

I see this just leads back to that ogre's tree traps. Even
though he's out of commission for the time being, thanks to
me, I'll still keep away from those vines. . . . I can see right
now there are no paths leading out of this forest. What kind
of a place is this anyway, with only one way of getting in and
out of every clearing? I could almost believe someone
planned it that way.

Hollow Tree and Fissure

There's the tree trunk and the fissure again. It looks like that may be my only hope. The cave is bound to lead to somewhere—and at least I won't be out in the open anymore. OK, here goes. I'll just tie this rope to the middle of the log . . . and climb . . . down.

Gee, it's hard to tell where the rope ends. I'm not sure. . . . Huh? What was that growling noise?! Ahhhhhh! What kind of creature is that?! I didn't see him before! I don't like his looks—that same look of hunger in his eyes. . . . Ahhhh! Stay away from me!

What am I going to do? I can't go back up! The rope stops here, and I'm too far from the ledge to jump. If I let go now, I'll fall into the abyss below. I don't have a choice—I'll have to swing myself. The only problem is that the closer I swing to *my* ledge, the closer I'll swing to *him!* But I have no choice—it's either do or die, and I don't want to die!

OK, here goes. I'll just . . . swing . . . as hard . . . as I can. . . . Come on. . . . That's it. . . . Y-y-yikes! Stay away from me. . . . Come on. . . . A little more. . . . Oh, no! He's starting to grab for me! I'll have to jump soon! OK, next swing. . . . I'll go for it. . . . OK. . . . Here goes . . . nothiiiing!

I made it! I made it! "Hey, stooge. Try to get me now. Next time leave your mouth closed, sewer-breath—it's enough to knock anybody out." I'm ready to ditch this joint.

Gee, this cave is pretty dark, and I don't have a flashlight with me—no matches, no lighter. Oh, for crying out loud, I can't believe that clown who was wearing my uniform didn't leave at least one book of matches in the pockets. But he didn't. Now how am I supposed to see in here?

Uh-oh. I just thought of something. What if that big oaf knows another way into this cave? Wh-what if there are more big oafs like him *living* in this cave? Oh, gee. No, I can't think about that now. I have to figure out how to find my way through this place. Surely nothing lives in a cave this dark. I can't see a foot in front of me. There must be some way to get some light in here.

Saaaay, what about my gem? Yeah, that's it! I'm surprised it doesn't glow through my pocket. I'll just whip that little beauty out, and—wow—it lights up like a Christmas tree. I lucked up in more ways than one when I found you, little jewel.

This is great. It works better than a flashlight. Now I can see exactly where I'm . . . whoa whoa whoa . . . goiiing! Sufferin' succotash, I fell through a hole! Where am I anyway? This definitely isn't the swamp area anymore, thank goodness.

Clearing After the Cave

My gemstone! Where is it? There it is. . . . Back in my pocket you go. You're much too valuable to lose. . . . Hey, it's the little pink fellows—*two* of them! What did he call me—a beanpole? "Hey, who are you calling beanpole? I'll have you know I work out. . . . Huh?" They want me to follow them. Fine with me—maybe they'll help me find my way out of here. "Hey, guys, wait for me!"

You should have taken a kinder, gentler approach when you met the first pink alien. Now they'll either help you or kill you.

So this is where those little pink creatures live! Wow! That guy up there on the rock—he must be their leader. It's a good thing I have this translator with me; otherwise, I wouldn't understand a word he says. "Boy, am I glad to see you, fellas! You're the only decent beings I've seen on this entire planet. I was hoping. . . . You will show me the way out? Oh, thank you so much! You don't know how much I appreciate. . . . Hey, where'd he go?"

Well, I guess those two standing next to that rock will show me the way out. "Before I go, I would like to say that I really appreciate your help. You're awfully nice to go out of your way to help little ol' me. Well, that's about it. OK, I'm ready to go now. . . . Guys? I said I'm ready to go now. Hey, guys, aren't you going to help me like your leader said?"

This doesn't make any sense. I don't understand what's going on. They said they would help me, and now those two are just staring into space like their brains have been put on hold.

"Hey, fellas, you said just say the word when. . . . Huh?" Now they're responding. These folks are really weird!

"I have to climb down in that hole? Are you sure there's no other way out? OK. Well, I'm not too crazy about this, but I guess I don't have a choice . . . for a being my size, right? Well, thanks again, guys. Hope we cross paths again someday. Bye!"

The Tunnel Maze

It's, uh, *dark* in here. Not just a little, either. It's *very dark* in here. Time to get the gemstone out again, but how am I going to hold it? I need both hands to climb down this ladder, and the passageway seems awfully narrow. I guess there's only one thing I can do—I'll carry it in my mouth. Boy, I feel like a dog bringing slippers to his master! I could get claustrophobic in here.

Wh-wh-what was th-that noise? That sent shivers all the way down my spine. Now I really feel like a dog—my hair's standing on end! I didn't like the sound of that at all. I knew this was too good to be true. How am I ever going to find my way out of here? And fast. I'm sure that . . . that *thing* I heard knows it isn't alone. Dollars to doughnuts it's carnivorous—like every other species on this planet.

Ouch! If I hit my head one more time! . . . I feel like I've been in here for hours—crawling and climbing, crawling and climbing. THERE HAS TO BE A WAY OUT OF HERE, OR I'M GOING TO SCREAM! Oops. I'd better be quiet. I don't want to draw the attention of that . . . whatever it is. I know this much—it must be pretty bad to live in a place like this. I can see it right now—I'll never find my way out of here. I'll die of suffocation, dehydration, starvation, or something, and my bones will just lie here and turn to dust. Either that or that *thing* will get. . . . A LIGHT! I see a light! Holy Moses, there really is a light at the end of this tunnel!

I'm almost there. . . . Just crawl a little more. . . . Yikes. I found the end of this tunnel, all right, but the pink aliens didn't say anything about crashing into a waterfall.

Waterfall Canyon

But how could I possibly complain? After the sorry sights
I've seen, this place is an oasis! I wouldn't believe it if I didn't
see it myself. Who would have thought this planet could be
capable of such splendor and beauty. It's breathtaking.

And I'm about to gag on my own words! Enough already,
Wilco; it's a nice place. I must be getting delirious. I've never
heard myself talk that way before. Maybe I was never inspired that way before. Even ol' Gladys didn't bring that kind
of poetry to the surface. Well, it's time to leave this lovely
place. I can see there's no alternative but for me to follow the
stream flowing through this watery canyon. At least that berry stench will be gone for good. I can't believe something
would actually want to eat me, smelling the way I did!

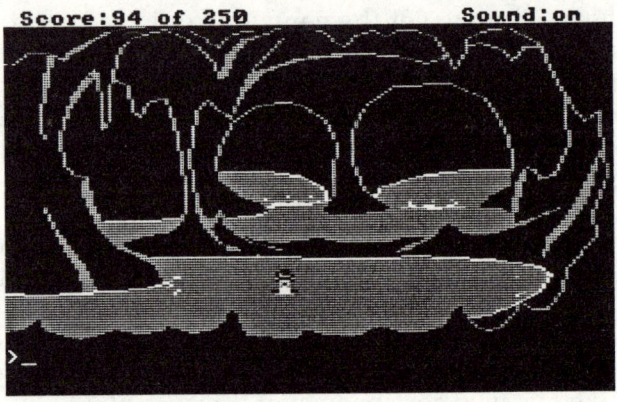

When reaching a fork in the watery canyon, you'd best take the road well traveled.

OK, this is going to be a little slow going, just like in the
swamp. . . . Well, what have we here? A fork in the water-
way . . . but which way should I go? I could always flip a coin
again. Nah, that leaves everything to chance. I have to get
into the nonjanitor mode I used when I was up against the
Sariens. Let's see; what would a nonjanitor do in this situation? I see two paths before me. One looks pretty rough, as
though it has seldom been traversed. Hmmm. And the other
path looks like it has a regular convoy passing through.

I don't exactly feel like taking risks right now. Well, I've
made my decision—I know which path I'm going to take.
Sorry, Robert—another time, another fork.

I hope I've made the right choice. . . . The pull is kind of
strong here. . . . On second thought, maybe I should have tak-

en ol' Robert's advice! There's a definite undercurrent in play now. . . . It's pulling me in circles . . . like a . . . bathtub . . . draaaaaaaaaain. . . .

Clearing Beyond the Watery Canyon

Hey, I'm out of the canyon! I can't believe I didn't drown in that whirlpool. I was really scared, but now I'm out in the open. I can actually see trees and forests again—and the light of day. I'll just swim over to the shore and find my way through that clearing. This has been a rather long day for me, and I'm not sure I can take any more surprises.

OK, now to find my way out of here. . . . There doesn't seem to be a path leading out of this clearing, either. What kind of place is this? Who can even get in here if there's no path? I can't walk through solid rock, and I sure don't have anything to cut through it. This is just dandy. I've come all this way only to be stumped once again by this idiotic terrain.

What else do I have in my pockets? I still have my gem—and this jockstrap and Cubix Rube. Gee, I'd forgotten about those two things. I was planning to turn them in to the boss after I finished cleaning out the shuttle—which never happened. The shuttle. That seems like days ago, now. Whew, I'm tired. I could easily lay my head down to rest right here.

But no, I have to carry on. I have to find a way out of here. Let's see; what else do I have in my pockets? . . . Oh, my Labion Terror Beast whistle. I've never even had the chance to use it. I wonder if there's really such a thing as a Terror Beast. Nah, no way. This thing's just for fun. It's probably just a glorified dog whistle or bird caller. Well, there's only one way to find out. . . .

Uh-oh, what's that noise? It sounds like rock cracking. Someone must be coming. Maybe I should hide. Let's see; where can I hide? . . . Huh? It's a . . . a . . . white tornado? No. It's a . . . real, live Labion Terror Beast. Oh, my gosh! I can't believe it! I didn't think they were real. He moves awfully fast—like a Tasmanian devil—much faster than I can move. He's headed straight for me—and look at that face. This guy's not fooling around. What can I do? Have I just whistled my own funeral march?!

Score:114 of 250   Sound:on

The Labion Terror Beast is as dumb as he looks. It doesn't take much to puzzle him.

"Uh, hi there. For your information I'm the one who whistled for you, but now I think it would be a good idea if you just left the way you came in. . . . Uh, tell you what—I'll give you . . . uh . . . this thing right here if you promise to go away and never come back. Believe me—these are very popular where I come from. It's only a matter of time before they'll be all the rage on Labion. . . . Well, here. Take it. Just don't take meeeee!"

Oooo, I can't look. . . . I know he's going to swallow me. . . . Huh? I don't believe it. That dumbbell can't take his eyes off the Cubix Rube. He's completely engrossed in it. This is my chance to get the heck *out* of here—through the hole he made in that rock. Strong little sucker. Just keep your eyes on that puzzle, Mr. Terror Beast, and you'll never miss me.

Boy, he did quite a number on this rock. I've never seen anything like it! And to think what he would have done to me. . . . *(gulp)* I think I'll take one of these rocks with me since I don't have any weapons. I may need to clunk someone in the head before it's all over. Now I'll just head the heck away from that guy. No more mail-order merchandise for me. . . .

Landing Platform

All right! What luck—a landing pad. And I was beginning to think I'd never get off this putrid planet. Uh-oh, another monkey to deal with. I'd hoped I'd seen the last of those creeps. . . . But nooooo. They're everywhere I go. Well, it's a good thing I picked up that rock on my way out of the clearing. I have a feeling it's going to come in handy now. But can I throw it that far? I have to hit him hard enough to knock him out. If I miss, he'll hear the rock, and I'll be as good as dead.

`Score:136 of 250`   `Sound:on`

Good shot. Your only alternative was to sneak by the guard, but then you wouldn't have had the opportunity to use that . . . um . . . sports thing.

Wait a minute. I can sling the rock at him. I'll use this jockstrap. Boy, am I glad I didn't give this stuff to the boss like I'd planned. One day I'll thank the clown who used my uniform for leaving all of his junk in my locker. He may have saved my life. OK, I remember how this is done. I used to sling rocks pretty well when I was a boy. Of course, those were only pebbles—this is the real thing. Well, here goes. . . .

All right. Got heem, Capteen! You overgrown baboon—guess you won't be swinging from trees anymore, huh? Heh-heh. Serves you right, jerk. I can't feel sorry for someone who works for a guy named Sludge. Now, what did you do with that weapon? It should be around here. . . . Oh, there it is. Let's see; maybe this thing didn't sustain too much damage. . . . Forget it. Cheap piece of garbage—I should have known better. They're no better than the Sariens when it comes to cutting corners.

Now to find my way onto that platform—this must be the elevator over here. Yeah, now we're in business. OK, what now? The darn door won't open. What could be the problem? . . . Wait a minute; there's a slot next to the elevator door. I'll bet I know what that's for! Smart thinking, Wilco, when you searched those overgrown rodents.

I knew it! It worked! I'm on my way up to the platform and soon to be OUT OF HERE—and it can't be soon enough for me. I've *had* it with this place! Sooo, here's the shuttle that brought me here—and it's the one that's going to take me home. *Home.* That's a word I cherish. I can't wait to get back to my closet.

Aboard the Shuttle

OK, I'm in the shuttle—now what? I'm *still* not an engineer, darn it. Maybe I should go back to school and study engineering. I thought sanitation management was a worthy and noble cause, but it's brought me neither the respect nor the buckazoids I want. Still, there aren't too many things that compare to a good custodian's closet.

Score:141 of 250   **Sound:on**

Most Xenonites fly imports. Unfortunately, yours is a domestic model. Don't be surprised if the gears work in reverse.

OK, what do all these controls and gadgets mean? Is there a manual in here? Shoot, I don't see one. I'll just give it my best shot and hope it's good enough.

Let's see. Power is probably a good start. Yeah, that started the engines. OK, now what? Altitude. . . . Boy, this thing's more complicated than I thought. Let's see. I guess I need to turn this dial in the vertical position. OK, that seems logical.

I guess the next thing is the Ascent Thruster. Hmmm, that's strange. It looks like the throttle here works backwards! Well, that figures. This must be a Xenon original—why the heck don't they stick to imports? They know these things are put together with rubber bands.

OK, throttle back to make this baby go up. It's moving. Oh, thank you, thank you. I'm on my way. . . . Now, what does that say? I've reached the proper altitude. Oh, I guess I should change the dial to horizontal now. OK . . . easy does it. Now what? It says the throttle works in regular mode now.

OK, throttle forward. That did it. Great! Now I'm really on my way.

I can't wait to get back to Xenon. I won't even mind seeing my jerky boss again. Heck, he's probably turning blue in the face right now, wondering where in the world I am. . . . At least I hope *someone's* noticed I'm missing. I just hope they believe my story, incredible as it's going to sound. . . . Wha? Huh? What is? . . . Oh, no! Not again! I don't believe it! I thought I'd seen the last of that . . . that slimeball!

"What do you want, Vohaul? Just get your filthy face off my screen! You lowdown, good-for-nothing mountain of fatback! You don't scare me! Nothing could scare me after what I've been through!"

Great. That jerk's taking me to his asteroid. Boy, he really has it in for me. Well, one thing's for sure; he's going to regret the day he ever saw Roger Wilco!

Game Points Earned

Action	Points
Choosing the right waterway at the fork	5
Climbing the hollow tree trunk	4
Diving into the deep part of the swamp	2
Finding and taking the gemstone	3
Getting a rock near the Terror Beast	2
Getting the gemstone after dropping it	1
Getting the keycard	3
Getting the key to unlock the cage	2
Getting the rope	2
Getting the spore	4
Getting the Terror Beast whistle	2
Helping the pink alien	5
Holding your breath before diving	2
Holding your breath before swimming back	2
Jumping from the rope to the ledge	5
Leaving the planet Labion	20
Mailing the whistle order form	2
Making it safely through the root monster	4
Making your way through the tunnel maze	20
Picking berries	4
Rubbing the berries on yourself	3
Safely hiding from the hovercraft	5
Saying the right word for the pink aliens	2
Slinging the rock at the guard	20
Swinging on the rope	2
Throwing the Cubix Rube at the Beast	10
Turning off the beeper	1
Tying the rope to the tree trunk	2
Using the gemstone in the big cave	2
Using the keycard at the elevator	5
Using the spore on the ogre hunter	5
Whistling for the Labion Terror Beast	5

Object Locations

Object	Location
Berries	Bushes behind the root monster
Cage key	Ogre hunter's body
Gemstone	Cavern beneath the swamp
Keycard	Dead guard at the crash site
Rock	Near the Terror Beast
Rope	On the rock by the cage
Spore	On the ground near the mailbox
Whistle	At the mailbox

Vohaul's Asteroid

Mission: Get rid of Vohaul, stop the clone infestation, and get off the asteroid.
Total Points: 84

Vohaul's Asteroid

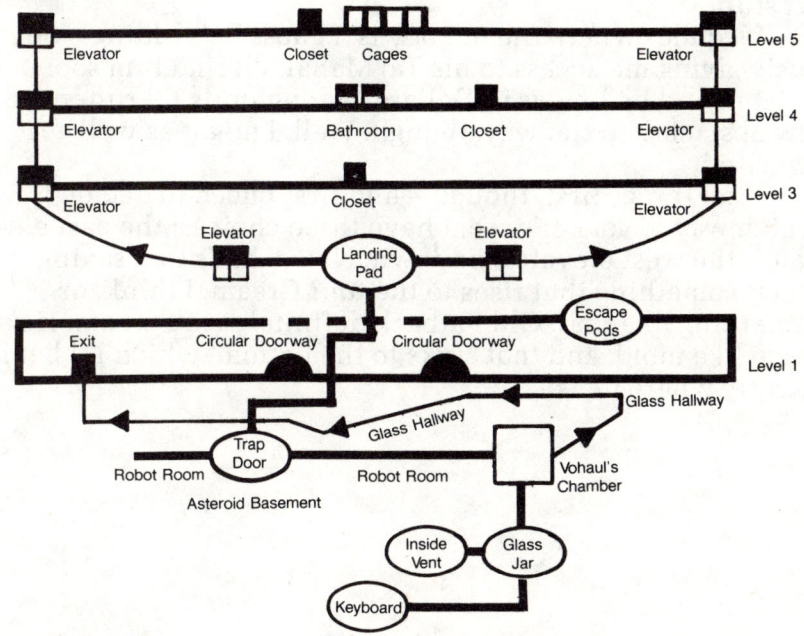

☐ Chapter 6

Inventory

---

glass cutter
lighter
oxygen mask
plunger
toilet paper
wastebasket

Asteroid Level 1

So this is where it all happens—the asteroid. I see there's no welcoming committee this time. I'm sure that tub of lard has much more interesting things in store for me.

My, this place is impressive, indeed. Who would have known the *enfant terrible* could turn a hollowed-out asteroid into this fancy piece of work? Suspended shuttle platform, circular entranceways, treadmill walkways—all state-of-the-art stuff.

I wonder where the old boy is. I guess he's hiding—obviously giving me access to his Taj Mahal. I'll find him sooner or later, and he knows it. Dollars to doughnuts I'll run into a few obstacles on the way, though. Well, I might as well get on with it.

First things first, though—and first I need to decide which way to go. Let's see; I have three choices: the east elevator, the west elevator, or down. Now what's that saying about something that rises to the top? Cream, I think, or something like that. Old Fatback definitely isn't cream. He's more like mold, and *that* sinks to the bottom, which I'll bet is exactly where he is.

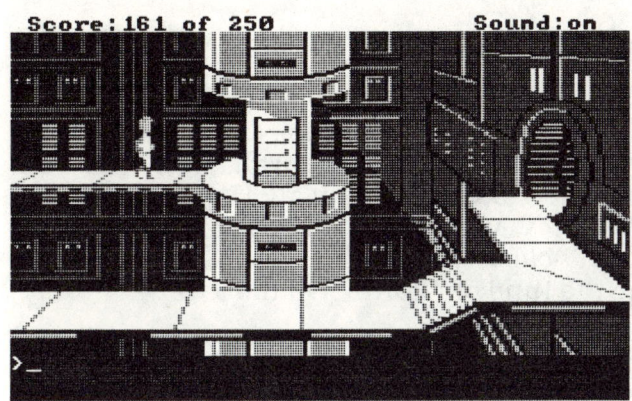
At this point, you'd be wise to go up, not down.

Keeping that in mind, I believe my choice is now east or west—so which elevator should I choose? Oh, what the heck—east is east, and west is west. What difference could it possibly make? I'll just use the easiest method I know to choose:

Eeny meeny miny mo
Tell me which way I should go!

OK, so it's east! Boy, this walkway sure is narrow. I'd better be careful and take it a little slower here. OK, now that I'm in the elevator, which floor should I choose? Heyyyy, there are only four buttons in here—1, 3, 4, and 5. I may not be good with numbers, but I *know* there should be a 2 in there somewhere. That dunce—he can't even count right. He thinks he has five levels, but there's no level 2. Hah, what a dope. Oh, well. I'll just try the next level up.

Asteroid Level 3

Ding, I'm here. It certainly works like a regular elevator—nothing strange about that. Hey, wait a minute. What the. . . . I took the east elevator, and now I'm in the west end of a hall. Well, whaddaya know? The twain did meet. Some architects *this* place had. Where'd they get their credentials, Tinker-

Toy U? Oh, well. Now that I'm here, I might as well walk east—there's nowhere else to go.

Gee, I have this real funny feeling I'm being watched. It's almost like a pair of eyes are following me. Oh, great—no wonder! What's this, "Candid Camera"? Well, I might as well smile and wave for the viewers. "Hey, y'all. Hope you enjoy the program." Imbeciles. Maybe they'd like some real entertainment . . . (clickety-clickety click-click-click). . . . "Hey, guys, this one's to the tune of "Puttin' On the Ritz"! . . .

> If you're bored
> And you don't know
> Where to go to,
> Why don't you go
> Jump off a cliff?
> Vohaul, you're the pits!

. . . (clickety-clickety click-click-click click-click-click clickety-clickety). . . . "It's time to say good night, Dick. 'Good night, Dick.' Y'all come back now, ya here?" (clickety-clickety click-click-click). . . .

Hope they enjoyed the performance. What zeros. They probably have cameras all over this zoo. That much I'm about to find out. All I know right now is that this hallway seems to go on forever. What kind of place are they running here, anyway?

Hey, wait a minute—a door. All right. Now I feel like I'm getting somewhere. What good does it do to be the suave and debonair detective-slash-secret-agent I am if there's no in-tri-gue . . . inter-gue . . . intreg. . . . Darn it, I can never pronounce that word! . . . If there's no *mystery* to resolve and nothing to detect? Darn it, now I can't even remember what I was doing. Where was I? . . . Oh, yeah. Now I remember—there's a door up ahead. Well, I'll just have to check this out.

Hmmm, the plot thickens. Vohaul's allowed me free reign of this ship—so far—and now I stumble across this little door. Well, we're about to find out what lies behind mystery door number 1. . . . Uh, but what if it's a zonk?! There's no t-telling wh-what could be behind that d-d-door.

I've got to look, though. I can't let my nerves get the best of me. I can do this. I can. OK, here goes. I'll just p-push this button and. . . . Ah! What is it?! Ahhhh! What's that thing on the floor over there!? . . . Oh. Ahem. It's, uh, a plunger. Well, uh, ahem. I wasn't *really* afraid. No one saw me, right? No, I didn't think so. I'm even out of the camera's range in here.

Heyyyy, this is a broom closet! Well, bless my soul; I don't believe it. There's certainly nothing to be afraid of in here. Gee, this reminds me of . . . home . . . *(sniff)*. Gosh golly darn. What I would give to be able to curl up in my favorite closet again—with coveralls for my blanket, a stack of toilet paper for my pillow. Oh, what . . . *(sniff)* . . . memories.

OK, Wilco, for the thousandth time, get a grip on yourself. First I'm scared, and then I'm emotional, when what I need to be is courageous! My gosh, if I fall apart over a broom closet, what's going to happen to me when I run into something *worth* falling apart over? This is ridiculous. I'll just take this plunger and be on my way. It may not be a real weapon, but it's better than nothing at all.

Well, this hallway's really hopping. Yeah, boy. There's so much excitement I don't know if I can contain myself. There's the elevator up ahead—I'll just make my way up to level 4. Uh-oh, there's another camera. I believe this Sludge character has a little voyeur problem. "Hey, guys, I'm back! . . . *(clickety-clickety-clickety)*. . . .

> Dressed up like a million dollar trooper,
> You look more like a dirty pooper scooper—
> A super-duper one!

(clickety-clickety click-click-click). . . . "Well, th-th-th-that's all, folks!" . . . *(click-click-click clickety-clickety)*. . . .

So old Sludge is watching everything I do. I'll bet he couldn't see me in that closet. I'm not as dumb as he thinks—that's all there is to it. I'll outsmart him yet. Now to try level 4. I'll just push the elevator button here, and away we go.

Asteroid Level 4

Dollars to doughnuts there's a camera outside this elevator door again. I *knew* it, darn it. I'm getting sick of this. Oh, well. Here we go again . . . *(click-click-click)*. . . . "Hey, fellas, aren't you getting a little tired of the same old song and dance? I know I am. But you know what they say—the show must go on!" . . . *(clickety-clickety click)*—Yeeeeooowwweee! What's that?! A *floorwaxer!* Leapin' lizards, and it's headed this way. I don't think this hallway's big enough for the two of us. Y-y-y-yikes! I'd better dash back into the elevator!

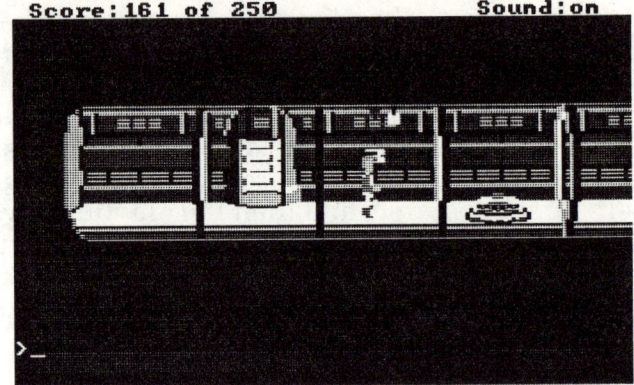

Score: 161 of 250   **Sound: on**

The floorwaxer. Sounds like something from a childhood nightmare. It's just a machine that waxes floors— and anything in its way.

Whew, that was close. That stupid floorwaxer—it sure took its time. Dumb robot. Hey, but wait a minute. That little guy did a great job on this floor. Wow, I can see my reflection! That stuff works better than Spic 'n Span, Pledge, and Mr. Clean combined. I'll have to find out what kind of wax he uses.

Oh . . . uh . . . that camera's still watching everything I do. "Yeah, I'll bet you guys think that was real funny. Sure, go ahead and make fun of me. We'll see who gets the last laugh around here. You're nothing but a string of sausages. Sueeey sueeey sueeey! Hey, Vohaul, you hear that? Someone's calling your name." Heh-heh-heh. Lousy jerks.

Well, after that last exciting excursion on level 3, I hope level 4 here isn't too ominous. Looks like just another run-of-the-mill asteroid hallway to me. Nothing much to look at here. Actually, I shouldn't be so casual about it. Who knows what Vohaul has in store for me? . . . Oh, wait! There's another door. I wonder what's in there. It's probably just another broom closet. No problem there, of course—it might be good for a few winks. I *am* getting a little tired, come to think of it. I've been at this hero stuff for quite a while now.

Now for the tricky part—pressing the button to open it. I got lucky with door number 1, but what if I'm not so lucky with door number 2? Gee, wh-what if the zonk's behind door number 2? I'd rather find Carol Merrill, but somehow I don't

think she's going to be there. Of course, she isn't going to be
there, Wilco; that's dumb. Why would Carol Merrill be all
the way out here behind door number 2? Oh, well. H-h-here
goes!

Ahhhhh! What's that?! What is it?! . . . Why am I screaming?! Ahem. OK, I'm not an idiot, and I'm not going to lose my
cool. It's just an empty closet except for that. . . . What *is* that
over there? Hmmm, it looks like a glass cutter. What in the
world would Vohaul want with a glass cutter? Let's see;
what does this tiny print here say? *U-Cut-It Glass Cutter, The
Ronko Corporation, Inc.* Oh, well. THAT figures. Xenon
closed its account with those jerks months ago. Of course, a
slimeball like Sludge would do regular business with them.
Ha, no wonder he has a useless glass cutter. There's no telling what other gadgets and thingamabobs they've unloaded
on him. Boy, how gullible can you get?

Yeah, I really need a glass cutter for oh so many things
and nearly every day of my life—just like that great U-De-
sign-It stud and rhinestone machine. That's the biggest piece
of junk I've ever seen. Like I'm really going to tacky up all
my uniforms with silver studs and colored rhinestones. It figures my jerk of a boss would give me such a useless birthday
present. Besides, the thing only came with eight rhinestones,
and of course, the boss was too cheap to order an extra supply for it.

There's obviously nothing else in this closet, so I'll just
be on my way. Well, on second thought, since I've picked up
everything else in my path, I might as well take this glass-
cutter thing, too. Who knows? I may need to cut myself out
of a glass cage—ha!

This is the darnedest excuse for a ship I've ever seen.
Miles of hallway, with nothing in between. And where's the
crew? Other than that waxer robot, I haven't seen another
soul. . . . Hey, finally. I don't believe it. There are *two* doors
up ahead. I wonder what that's all about. I'll just check them
out one at a time. Monty never said anything about door
number 4.

Well, I certainly recognize *this* place—I think I've logged
enough hours cleaning bathrooms. And I see the other door
leads in here also. I'd say by the look—and smell—of things,
no one has logged *any* hours in here—EVER! Ooooh, I don't
know if I can stand it. Look at this floor! The hallways are

spotless, and then you come in here, and it's like walking in a sewage tank. Fatback must spend about 90 percent of his time in here—it has his aroma.

Who are all these weirdos in here? I have yet to see the first crew member walking a hall, and then I come in here, and there's only one stall empty. Well, I could always ask. . . .

"Hey, you in there, where is everybody? Why aren't you and the rest of the crew keeping guard out in the halls like you're supposed to? . . . What's the matter—you don't speak English?" Great. Another closed-mouth jerk. "You know, some castor oil might help that little problem of yours. And get a can of Lysol, pal—that's what it's for."

I'll try someone else. "Knock knock, anyone in there? . . . So what are you going to do about it, bozo? I'll stand here all day if I want. If you ask me, you're pretty rude for taking so much time in there. How do you know there isn't a line out here waiting to get in? Maybe there's a line stretching all the way down to level 3—in both directions. Yeah, that's it. And we've been waiting for three hours. . . . Try chocolated Phillips next time, turkey."

Good gosh, what in the world do they eat on this asteroid? Probably garbage. This is the grimiest place I've ever seen. And what are those *facilities* over there? I'd hate to be an eyewitness to one of those things in use—gee! I feel dirty just being in this place, and naturally, there are no paper towels. That's typical. I'll just check out that empty stall and see if there's an extra roll of—bingo! I don't believe it. I'll just snatch up this roll of toilet paper while I have the chance. I'm not getting stuck without paper towels again. Hey, what's that scribble on the wall—all over the wall? Alien graffiti. Give me a break. Bunch of weirdos.

Hey, wait a minute. Since I'm here, I might as well leave a few scratchings of my own for the scrutiny of mankind . . . er . . . sludgekind. Let's see; what should I write? It needs to be something profound—something memorable. I've got it! *Roger Wilco was here. Roger 'n' Gladys 4-ever. Roger + Gladys = Luv.* There, that should do it. That wall could be a national monument someday—in my honor.

Whew, I can't take this much longer. I'm going to have to make an exit from this room—el pronto—before I die of odor abuse. But first, one quick check in the mirror won't hurt. Well, if I must say so myself, I look pretty good, considering what I've been through. The uniform's a little dirty, and I could stand a dollop of mousse in my hair—not to mention a shave. Nah, on second thought, I look pretty darn suave and debonair with this three o'clock . . . one o'clock . . . whatever it is, shadow. Señor Johnson, eat your heart out.

Boy, I'm glad to be out of *that* place! Now I feel like a walking bathroom. That stench is still in my nose—ugh. I'd like to get the heck off this floor. Sludge Vohaul is even grosser than I thought. Great, there's the elevator up ahead— oh, no—and that stupid camera again.

One more time . . . *(clickety-clickety click-click-click)*. . . . "Eh, what's up, doc? Hey, Vohaul, too bad about your ship's library going up in flames and all. Yeah, I was real sorry to hear about that. That was a real shame, having all two of your books burn. And I understand you hadn't even finished coloring in one of them. Nyak nyak nyak. Later, chumps." . . . *(clickety-clickety click-click-click)*. . . .

Boy, I'll bet those guys are really getting mad at me now. I just can't figure out why everything's been easy so far. I'd better not get too confident—I'm sure the worst is yet to come. One more level to explore—level 5. This sure is a large asteroid to have only four levels—well, five, counting the basement, which I'm sure is where Vohaul's hiding.

Asteroid Level 5

OK, here I am on level 5. I'm running out of show-biz routines for those darn cameras, though. But gee, I can't just *walk* by one without doing something—how boring! No, I have to think of something before I step off this elevator. Let's see. . . . Oh, I've got it! "Hey, guys!" . . . *(clickety click-click-click)*. . . .

> Three-six-nine
> The goose drank wine
> The monkey chewed tobacco on the streetcar line
> The line broke
> The monkey got choked
> And they all went to heaven in a little rowboat!

Three-six-five
I'm still alive
You think you have me worried with your gutter jive
But talk is cheap
Still sewers run deep
Your mama is a cesspool, and your daddy's a creep!

That ought to get him going for a while. . . . "Au revoir, les misérables." . . . (clickety-clickety click). . . . I just hope I'm not playing with fire here. I might be getting a little too bold for my britches. I still can't get over how everything's been so quiet—too quiet. Something's bound to break soon. I just know he's going to try something. I only wish I knew what—and when.

Hmmm, this is strange. Those look like cages down there at the east end of the hall. Maybe that's what he has in mind for me. He's going to sic one of his scary monsters on me as soon as I walk by those cages! Or maybe he's going to release all of them! Y-y-yikes, I even see some pretty big paws wrapped around the bars of one of them! I don't know what kind of creature that is, but his hands are as big as my head! One thump from that guy and . . . (gulp) . . . it's curtains for me!

What should I do!? Saaaay, there's a door over there between two of those cages. Well, how conveeeenient for it to be *right* between two cages. I'm sure that wasn't planned or anything. I have to know what's behind that door. There could be something vitally important to my mission in there! Maybe if I huddle close to the outer wall, those . . . creatures . . . whatever they are, won't be able to see me. One way or another, I just *have* to get in that door.

Gee, James Bond wouldn't act this way. But then again, I'm no. . . . Oh, to heck with that broken record. Even Dr. Zachary Smith wouldn't be such a ninny about this. OK, I'm going to do it. I'll just walk . . . very . . . slowly . . . up . . . to . . . the door. I did it! I'm afraid to look, though. I don't see anything reaching out to grab me. I can hear my heart beating in my ears again. It sure is loud. Now I just have to push this button and go in. Don't even think about zonks behind *this* door, Wilco. I've got enough to worry about without thinking about that.

I'm in. Whew, that was scary. Maybe I was wrong. Maybe I had nothing to fear but fear itself. Sure, that's what it was. I'll remember that next time. I'll be much braver. Now, to see where I am. Well, that figures—another dang closet. I've been through three levels of this asteroid so far, and the most I've seen is a few closets and some cages. Don't these people *do* anything around here?

This dumb hole in the wall isn't even big enough for a good snooze. What's that? A pair of cruddy old overalls. Yuck. They stink to high heaven. I don't need that rag. Vohaul probably wore it when he was 20 sizes smaller—and he's obviously never heard of using a washing machine.

What was that noise? Something just fell out of the overalls. Hmmm, a lighter. Now this little thing could really come in handy. I could appreciate lighting a few fires under Vohaul's. . . . I'll just drop this little baby in my pocket. Hey, my pockets are getting too full. How am I supposed to carry all this stuff?

There's a wastebasket in the corner. Maybe I can use it to haul around some of this junk I've collected. I'm beginning to feel like a regular pack rat here. At least I can carry everything more easily now. This uniform doesn't exactly have unlimited pocket space.

Now, I'll just carefully make my way out to the other side of the hallway, away from these cages. Maybe I could take a little peek in one of them—just a little peek—but not in the one with the gorilla paws. OK, I'll be swift about this. . . . Hmmm, I see something moving around in there, but it's so dark I can't tell what it is. Oh, well. Looks like this was no big deal after all. Still, I think I'm better off walking near the outer wall.

On to the other end of this hallway now, where I'm sure there's another camera waiting for me. . . . Huh! What's that?! That cage—I can see inside it from here. OH, NO! It's a Black Spiny Kisser—at one time the most dreaded beast on Xenon! But those creatures—I thought they were extinct. Darn it! They're not extinct—they just moved on to another planet, and now Vohaul has one for a pet!

Jeepers creepers, they're the ugliest darn things, too— skinny as rails and with the biggest, fattest, slobberiest lips on any creature alive! And they love humans! One kiss from the dreaded Black Spiny Kisser is worse than *ten* stings from a scorpazoid. Oh, no. The door—it's opening. I can't let that thing see me! One look at me and it'll never give me a moment's peace! Some of them can even walk through walls!

Y-y-y-yikes! I'd better go back the way I came. I can't let that creature catch up with me! Ahhhhhhhh! . . . I can make it. . . . Just run as fast as you can, Wilco. . . . Forget the camera—who *cares* if they make fun of me this time?! I don't give a flying fig. . . . Just get me out of here. . . . Almost there. . . . I just have to jump in the elevator and push the button!

I did it! Boy! My heart's racing so fast I wouldn't be surprised to see it leap out of my chest and run away! Ugh, what am I saying? What a horrid thought! I'm going to have nightmares about that Black Spiny Kisser for days to come. So that's what Vohaul thought he'd do to me—sic his wretched, love-starved pet on me!

Well, I'm sure they all got a good laugh watching me dash into the elevator like that. So what. He has to know this little game isn't over yet. The laugh's going to be on you, Vohaul. You just wait. So I'm back to level 1 now. Looks like there's only one way to go from here—and I'm sure the proverbial gates will be swinging wide open for me.

Asteroid Basement

OK, back over these narrow walkways and to the steps leading down to what I know must be Vohaul's lair. This *is* an odd layout. What a waste of a lot of neat state-of-the-art technology. It figures old Fatback wouldn't know how to put it to good use—he and his polyester insurance nerds.

OK, now that I'm at the bottom of the stairway, which way should I . . . huh?! How did that happen? OK, I'll just pick any direct. . . . Huh! Now wait a minute. Oh, no. I'd better make a dash for the . . . ouch! That darn wall practically came down on my head! Oh, geez. I'm trapped! Now what do I do? I can't believe that sorry. . . . Uh-oh, what's happening to the floor? It's moving! Oh, no! . . . What's that stuff beneath it? . . . It's *acid!*

OH, NO! I'm a goner for sure now! What the heck am I going to do?! I have to think—fast! That floor is moving slowly, but it's only a matter of time before it opens all the way and I'll be sucked into a hole of bubbling acid! . . . *(gulp)* . . . I don't even want to *think* of what that's going to do to my complexion.

I have to think of a way out of here. There must be a way to escape. These walls are too smooth to climb. The ceiling is nothing but a bunch of circuits and light fixtures. If I had something to throw up there, I could swing from one of the overhead beams or lights—but I don't have anything to throw. What about my clothes? Oh, this suit isn't long enough to reach. Those beams must be 20 feet above the floor. The floor's almost open! All I can do is crouch against the wall, but soon there won't be any floor left for me to stand on!

I can't believe this. After all this time, I've survived every obstacle thrown in my path, and now he gets me with a trap door—the oldest trick in the book. And all I have at my disposal is a bunch of bathroom stuff—a roll of toilet paper, a plunger, a . . . PLUNGER. That's it! Maybe that will work! We used to use these things to walk all over walls and ceilings when I was a kid! That was when I first thought about being a janitor.

OK, my funny-looking friend, please do a good job for me now . . . because this floor . . . is almost . . . GONE! It's gone! Here I am over a pool of acid, hanging onto a plunger, of all things, for dear life! I just hope my strength doesn't give out. I just hope this *plunger* doesn't give out.

The floor—it's closed! I'm safe! Yahoooo! The trap doors are gone! I did it! I made it! I survived again! I'll bet that mountain of lard thinks I'm nothing but a memory. I fooled you again, sucker! Nah-nah-nah-nah-nah! Suckerrrrr! Who the heck does he think he is anyway?

Whew! That was close, indeed! I'm just about worn out, but I have to carry on. If nothing else, I just have to find a way off this ship so I can warn everyone at Xenon about the impending clone infestation. They'll never believe it. Now which way should I go—east or west? Let's see; my mind is just too confused to make a decision right now. I've already used eeny meeny miny mo, so that won't do again. I've got it!

> Engine engine number nine
> Going down Chicago line
> If the train should jump the tracks
> Will you go forward or back?

Aha, it's west this time. Well, that's nice for a change—I hope. I'll be very careful this time. I don't want to walk blindly into any traps again. Let's see; what's in here? Hmmm, a wall of *red* robots, and I mean RED. What are they, mechanical fire fighters or something? Uh-oh. One of them is moving. Uh, it's coming this way. It may be coming at a snail's pace, but it's still coming this way! I'd better dash back out into the hall.

No trap door again this time, thank goodness. Uh-oh, it's still coming. What am I going to do? He sure is moving slowly—obviously not a firefighter. But I have to think fast! He's going to catch up with me sooner or later—I can't just stand here! I wish I had a bucket of water to throw on that bucket of bolts—that would rust out his joints for good.

Hey, wait a minute. Those circuits up there—they lead into the robot room, and I'll bet they're connected to the wall units holding the robots. What was it we did to short-circuit those robots that went haywire back in my sixth-grade electronics class? Oh, I remember! We overheated the circuits! That's it! Those little things were going nuts all over the room! I have this lighter, so maybe I can get those wires hot enough to short-circuit that creep and all his buddies, too.

Oh, shoot! I can't do that from here. The ceiling's way too high, and the only thing I have to stand on is this wastebasket. Wait a minute! I have paper to burn! I could burn my roll of toilet paper, and maybe the flames would be hot enough to reach those wires! But I can't just hold it and burn it—I'll end up burning off my hand. I have to put it in something. . . . But of course! Ding! The wastebasket! I'll put the toilet paper *in* the wastebasket and set it on fire!

Jiminy crickets, that walking cola can is almost here. I don't have a second to waste. . . . There! Oh, please! Burn, baby, burn! That's it. . . . Yikes, get away from me! Ahhhhh! Don't come near me, you rusted-out canister of aluminum and . . . huh? He stopped! It worked! He's dead as a doornail—whatever that's supposed to mean!

Yahooo! I foiled the odds again! Thank goodness I was smart enough to collect all this junk. We heroes just never know when some seemingly insignificant little doodad might come in handy. In this case, that toilet paper saved my life! Now what should I do? I didn't see any doors leading out of the robot room, so this time I'll go east. Hmmpf. See if I trust that little engine number again.

What? Not again! Another room of red robots? I don't believe it. But they're all shorted out, thank heavens. That darn
Sludge—he would have had me no matter which way I went
this time! I do see a door on the far side of this room, so I'll
just make my way over there and through it—as if I really
have a choice.

Vohaul's Chamber

"Vohaul! You lousy, good-for-nothing surplus of flesh. I
knew you'd be down here somewhere. I always heard sludge
sinks to the bottom. So how'd you like my performances? I
thought they were all pretty good, if I say so myself. You
really inspired me to come up with some great material. I'm
sure your troops enjoyed them as well. So what do you
think? Will I have a decent shot at 'Star Search'?"

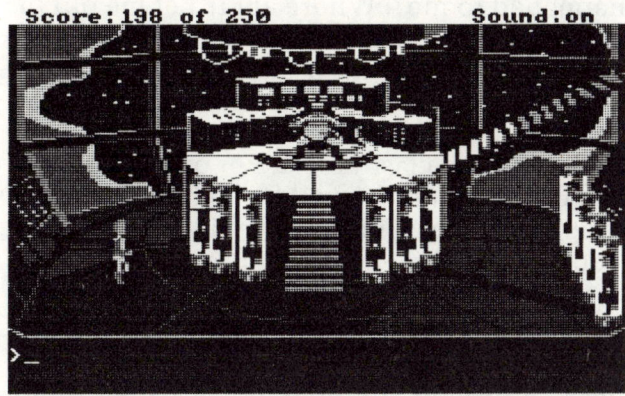

Score: 198 of 250   Sound: on

*There he is again,
in all his grossness,
sitting in that same
position, surround-
ed by all those
nerdy clones.*

"You think I'm afraid of you and your fleet of nerds?
Well, I've got news for you, Fatback. You'll never get away
with your disgusting plan to infest Xenon. You mountainous
oaf—you can't even get up out of that chair. What's the matter—afraid the whole asteroid will tip over? Well, I've noticed that it's extra large. Ha-ha-ha.

"Oh, really? Is that a fact? Yeah, well, that and a dime
will get you a cup of coffee. . . . Oh, no kidding. Thanks for
the blinding glimpse of the obvious. You look like one big
experiment.

"I'll show you just how bold I am, Mt. Jumbo. I'll walk
right up to your platform—it won't cave in now, will it?—
and look you right in your ugly eyes, and . . . hey! What's

109

happening?! No fair! What's going on? I can't see a thing!
What are you doing to me, you 2000-pound pile of dung?"

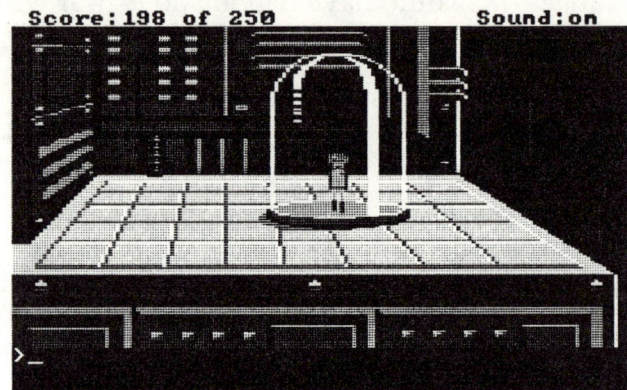

*Janitor in a drum?
If you can't figure
your way out of this
one, you might as
well cut your losses
and quit.*

"Hey, what happened to me?! Where am I? Let me out of
here! This is a travesty of justice! You'll never get away with
this, Vohaul!" I don't believe it—I *am* in a glass cage. That
creep reduced me to tabletop size! How the heck am I ever
going to get out of this mess? Well, one thing's for sure. I still
have my handy-dandy Ronko glass cutter. For once, I'm
thankful the Ronko Corporation exists. Maybe I'll even per-
suade Xenon to reopen our account with them—assuming I
make it back, of course.

I don't have a moment to lose. That infestation is slated
to start any minute now. I'll just cut my way out of here. . . .
Good thing he's so big and fat; he doesn't even have the mo-
bility to turn around and watch what I'm doing. Good thing,
too, none of his cronies are hanging around here. All right,
I'm out! Now what to do?

I see a vent over there. Maybe I can find a way to short-
circuit some of *his* wires. I'll just crawl in through these
openings here and. . . . Oh, gee. Would you look at this place?
It's like . . . mechanical intestines or something. Yuck. This
must be Vohaul's life-support system. And that button over
there—I'll bet it's the key. Say your prayers, Porky—not that
they'll do *you* any good—you're about to pollute the universe with your stench for the last time.

I did it! It's done! What was that he said? The clone infestation is still set to go, and now he's rigged the asteroid to
blow up! Oh, great. I've got to get out of here—now! I must
find a way to bring myself back to normal size; otherwise, I'll

blow up with the rest of these schmucks, and Xenon won't stand a chance.

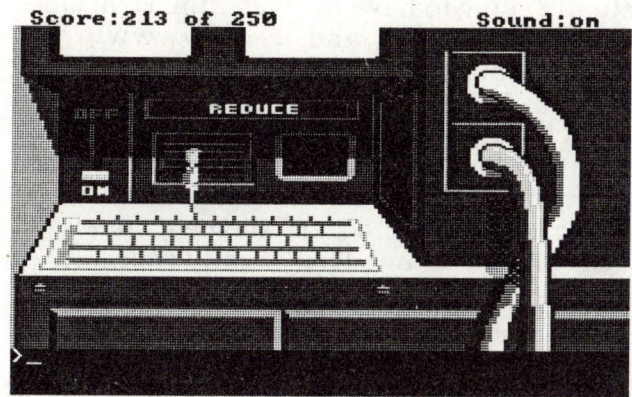

What did we tell you? Heroes always draw on their resources. Good thing you took that typing course 20 years ago.

OK, I'm out of Vohaul's innards. He's dead. At least I won't have to worry about that creep anymore. Now what? What did he do to reduce me this way? I feel like I've been through a color photocopier. I'd better be careful making my way around this machinery—I wouldn't want to fall off. What a wimpy way to die that would be. Hey, a computer terminal! And there's a lever over there that turns something on and off. I'll bet it's the reducer mechanism. . . . OK, Wilco, gather up all your strength and *pull* that lever as hard as you can. Oooh . . . oh . . . it's heavy . . . but I did it!

That monitor is running a message now. It says *Reduce or Enlarge?* I know what I have to do now. I may not be James Bond, but I can sure as heck type—and I do windows, too. So here goes . . . E-N-L-A-R-G-E. I hear a strange noise. It must be the enlarge mechanism firing up. Now I have to hurry back over to the glass cage so I can be returned to my normal size. Oops! I'd better be careful. This place is slippery, and I *definitely* don't want to fall off! Great, it's working. . . .

Hallelujah! I'm myself again! I'm my normal size! I'm so smart for figuring this out I could kiss myself! I deserve a reward! . . . But the game's not over yet—no way. I still have to figure out a way to stop that clone infestation. What's that monitor over there in front of Vohaul's chair? Let's see; what does it say? Oh, gee! It's the terminal he used to activate the clones! I have to type in a code to stop it from happening! But what code? Where am I going to find a code?

Oh, no. Ugh. I'm going to have to search *it!* He must have the code on him somewhere. It's my only hope—there's no alternative. I'm actually going to have to *touch* him. Ugh! Oh, well. I'll hold my nose, grit my teeth, and . . . eeewwww! He's even grosser than I thought. He really is the scum of the earth . . . er, the asteroid. This is sickening. Come on, you overgrown toadstool. I know that code's here somewhere. What's that written on his hand? Oh, he's such a nerd. Even *I* don't write stuff on my hands! What in the world does that mean—*SHSR?* Shoozer? What's a shoozer? Maybe it stands for something. . . . Gee, I don't know. That's really weird. It must be a code for some. . . . CODE?! That's it! I'll bet it's the code to stop the clones!

Ohhhhh, what are you wasting time for, Wilco, you dim-wit? I have to type in these letters correctly. Let's see; *S-H-S-H* . . . oops . . . *R.* There, that's it. That should do it, I hope! . . . It worked! The screen now says the clone infestation has been aborted. Yeah! I did it! I saved Xenon once again! And they'd sure as heck better be thankful for it, too. I've had enough of this second-class citizen treatment. And my days on Orbital Station 4 are OVER!

Now I'd better get the heck off this asteroid. I have no idea where I'm going, but I'm bound to run into some sort of shuttle bay sooner or later—hopefully sooner. Sure, I'd like to have my name up in lights, but I don't want to *be* those lights! Well, I know where the other doorway leads; I might as well try the staircase now. Boy, this thing is narrow. Everything in this ship is narrow—except old Sludge. Talk about having a false image of yourself! No way he could have walked up these steps—the whole staircase would have fall-en through.

Glass Hallway

Now where am I? What kind of hallway is this, made of pure glass? It looks pretty fragile. I'm not sure I trust this thing. What if the glass breaks? Then there wouldn't be anything left between me and the rest of the universe! I'm not sure I like this, but it looks like I don't have a choice.

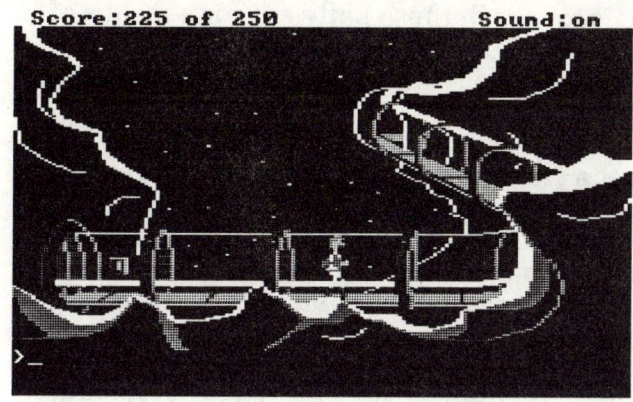

You might want to take precautions here; glass is fragile, you know. You don't want to breathe without oxygen.

Say, what's that thing on the wall? Hmmm, it looks like some sort of emergency box. I'll just take a peek inside and see. . . . Hey, it's an oxygen mask. Why would there be an oxygen mask in here if this glass hallway were built to last? Shoot, this glass is probably just as cheap as that stuff they used for the windshield on the *Arcada* escape pod. Gee, that seems like centuries ago.

Still, I'm not taking any chances. I'm going to wear this mask—I don't care how silly it looks. Who's going to see me anyway? Better safe than sorry if this thing cracks. I wonder where this tube leads to anyway. Guess there's only one way to find out.

This thing is like a maze, zigzagging all over the place. Yikes, the glass *is* cracking! I knew it was cheap! There's even a hole in it now! Oh, my gosh. I've got to get out of here—she's breaking up; she's breaking up! There's a door up ahead! I'll make a run for it! . . . Whew, I made it—once more! I *knew* that glass wouldn't hold up. It's a good thing I found that mask and was smart enough to wear it.

Level 1 Hallway

Accck! That noise—those sirens—they're killing my ears. It's true then; the destruction of the asteroid is imminent, and there's not a darn thing I can do about it. It looks like I'm back on level 1 now—only I haven't been in this hallway before. Of course not! This leads to one of the circular doorways—the west one, which means I'm in the west wing. I've *got* to find a way off this ill-fated rock.

My only hope is to search these hallways for a means of escape, but which way should I go? Should I choose east or west this time? No eeny meeny miny mos or engine engines are going to help with this dilemma. If I choose the wrong way, it could cut into my already-limited time. Since going east leads back to the circular doorway, maybe I should try west. If I go through that doorway, I know I'll have to cross that very thin, suspended walkway—I'd be doing a real balancing act, walking on that.

I'll go west. Something tells me, though, that the east hallway and the west hallway are one and the same—after all, I am inside a sphere of sorts. So once again, the twain will meet. How clever of you, Sludge. Oh, well. There's no point wasting any more time—especially since I don't *have* any more time.

I wasn't kidding when I thought this hallway stretched into oblivion. . . . Wait a minute! What's that up ahead? . . . Could it be? . . . Are they? . . . They are! Escape pods! Bingo! I *knew* there had to be a way off this asteroid! Uh-oh, what's that next to the? . . . Great balls of fire, it's a biped! Y-y-y-yikes!!! G-g-get me outta here! One touch from that two-footed tin can and no more Roger Wilco!

That darn thing's going to follow me all the way to the circular doorway. There's no way it can make it through the suspended walkways, though—not on those clumsy feet! I'm sure I'll be safe once I reach the doorway . . . if only I would hurry up and get there. Yep, it's still behind me all right. . . . Almost there . . . almost there . . . way to go! I made it! And that ding-a-ling did just what I thought—he turned around and went back to watch guard over the escape pods.

I have to figure out a way to get into one of those pods. I can see right now the west wing isn't going to do me any good. Every time I go near the pods, that biped will be after my hide. Heyyy, but he stands on the *east* end of the pod room. Maybe if I come in through the *west* end, I'll have enough time to get in a pod before he can reach me! That's it! That's my only hope!

Now to make my way down the suspended walkway . . . narrow as it is . . . and to the east circular doorway. . . . Great. Now that I'm here, I'll just walk as quickly as I can. I'm running out of time, but I'm bound to run into that escape room soon. . . . Great, there it is, just ahead of me. Now, I've got to be careful about this. I know the biped can't move very quickly, but still, he may be able to move fast enough to reach me before I can get in the pod. And if that happens . . . *(gulp)* . . . well, I just won't let it happen. I can outrun that piece of rubbish any old day. I've already proved that—haven't I? Oh, gee. Well, if nothing else, maybe I'll have enough time to at least open the pod door.

W-well, here g-goes n-nothing. Just keep your distance from me, you . . . y-y-yikes! He sees me! He's coming! I have to push the button to open the pod door. . . . Did it! Yahooo! "Run, run as fast as you can, you aluminum refuse. You can't catch me; I'm the Roger Wilco man! Nah-nah-nah-nah-nah." Wow, that was easier than I thought. What in the world have I been afraid of? I could stop and read a novel before that clown catches me. . . . Uh, but just in case, I'll keep going. Wouldn't want to take any unnecessary chances now that I'm almost home free.

Well, I made it back to the circular doorway once more, and just as I thought, old flathead back there didn't bother to follow me through. He's on his way back to guard the escape pods by now, so I'll just be on my way back there—it'll give the poor guy something to do before he's blown to smithereens.

Darn hallway—it stretches on forever—darn stupidest design. . . . Thank goodness I'm almost there. . . . One more hallway to pass through before the great escape. . . . OK, I've arrived! Now I just have to do this as quickly and cleanly as possible. No slipping or sliding or falling down allowed. Fumble this one, Wilco, and you can kiss old Gladys and the rest of the universe good-bye. Gladys—what I would give to see her lovely face right now—even if it does have a communications board attached to it.

Well, here goes. . . . "Hey, sucker, I'm outta here!" Y-y-y-yikes!! . . . I'm in! I made it! I really did it! Roger Wilco is BACK IN BUSINESS!!! Now, to get this baby rolling. Oh, heck—no way these pods were made for Vohaul's escape. That lard bucket couldn't fit in ten of them put together. He must have had an extra-large shuttle docked here some-where. Ha, it probably takes up the rest of the asteroid.

Let's see; I really have no time to waste now. There's a button labeled Launch. That must be it. . . . I'll just push it and . . . hallelujah! I really am out of here! Way to go, Wilco!

"Eat dirt, schmuckos! You're gonna go out with a real BANG!"

Well, once again I've saved Xenon and the rest of the universe from the hands of some slimy, lowdown, good-for-nothing creep!

Aboard the Escape Pod

Ahhh, peace and quiet at last! Just me, the pod, and a thousand points of light. I never dreamed life could be so . . . so . . . calm. Not a worry in the world. I'm just drifting through space. This is wonderful—this is the life. I've certainly earned this solitude.

Oh, and the utter contentment I feel, knowing the universe is safe once more, all because of . . . huh?! What's that noise? Don't tell me this thing's going to start making a bunch of racket just when I was getting used to all the. . . . What? I'm running out of oxygen?! I don't believe it! No, this can't be! Vohaul, you lousy scum, how can you expect to escape without oxygen? Of all the dirty, rotten. . . . Saaay, that looks like a sleep chamber over there. Hey, no problem. Maybe old Vohaul wasn't so dumb after all.

I'll just crawl into this comfy little sleeper—all safe and sound—and dream away. Oh, I've been wanting a good snooze . . . (yawn) . . . I wonder what Gladys and Fly-Boy are up to. . . . Fly-Boy . . . haven't seen him since they put me on . . . (yawn) . . . Orbital Station . . . 4. . . .

Get ready for a long winter's nap. By George, you've earned one!

Game Points Earned

Action	Points
Dropping the wastebasket	1
Entering the escape pod	10
Entering the pod sleep chamber	10
Escaping the acid pool	10
Escaping the glass jar	5
Getting the glass cutter	1
Getting the lighter	1
Getting the oxygen mask	2
Getting the plunger	1
Getting the toilet paper	1
Getting the wastebasket	1
Putting the toilet paper in the basket	1
Short-circuiting the wall of red robots	10
Terminating the clone launch	10
Terminating Vohaul	10
Using the plunger	10

Object Locations

Object	Location
Glass cutter	Level 4 closet
Lighter	Level 5 closet
Oxygen mask	Glass hallway
Plunger	Level 3 closet
Toilet paper	Level 4 bathroom
Wastebasket	Level 5 closet

Chapter 7
The Pirates of Pestulon:
St. Elmo's Ire

OK, Wilco, it's time to wake up and smell the coffee . . .
er . . . garbage. That's right; it's time to rekindle those old adventure fires that once burned so bright. You know, the ones
that had you on such a roll the last time we saw you. If memory serves us correctly, you managed to cough up enough
chutzpa to strike a final blow to Sludge Vohaul's plans and
put him out of commission forever. Pretty impressive, considering you were flying solo again.

Solo seems to suit you, though, so now it's time to get out
the old bellows and start fanning those flames again. But be
prepared—you aren't really going to find all that much adventure here in your *immediate* surroundings—more like a
few headaches. You see, while you were fast asleep, your little ship was pulled in by none other than a robot trash barge.
That's right—a ship carrying pure, unadulterated *garbage*.
Not exactly what you were hoping for, huh?

Well, it looks like you're going to have to forget about
Gladys, Fly-Boy, and the rest. Your only priority at the moment is to find your way out of what continues to be a very
smelly predicament. Of course, if you manage to accomplish
such a feat, you're in for some real surprises—and we're not
talking about the elevator ride at World O' Wonders, either.

On second thought, maybe you *should* forget about leaving the trash freighter. Life, unfortunately, isn't going to get
much better for you whether you leave or stay. Of course,
there is one saving grace—if you do leave, you'll finally get
to sink your teeth into one of those big, fat, juicy Monolith
burgers. That's about all you have to look forward to—la-
mentable, but true. Sorry, pal. Your lucky star must be stuck
below the horizon.

The Trash Freighter

Current Goal: Find your way off the trash freighter and reach the planet Phleebhut.
Total Points: 145

The Trash Freighter

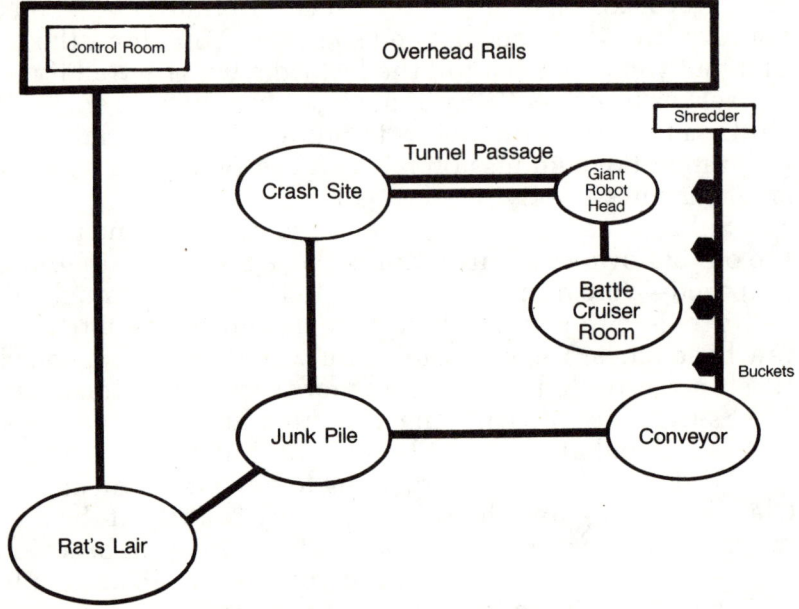

Inventory

buckazoids
gemstone
ladder
old wire
power reactor
warp motivator

Trashed Pod Room

(*Yawn*) . . . ohhhh . . . (*yawn*) . . . huh? . . . Ohhhh, I feel like I've been asleep for a thousand years. "OK, boss, I'll get right . . . (*yawn*) . . . to it. Can't a guy get any slee— Huh?! Where am I? What happened?" Oh, no. Now I remember. Vohaul! The sleep chamber! He's dead. I almost ran out of oxygen. Now I *do* remember. I'm in the escape pod from Vohaul's dearly departed asteroid!

I really might have been asleep for a thousand years! There's no telling what the universe is like now. I won't know a soul. Everything will be different. Orbital Station 4 might not even exist anymore. That means Gladys . . . and Fly-Boy . . . they've been dead for over a *thousand* years. I really do feel alone now. Do I dare step outside the safety and security of this escape pod? They won't know me. No one will know me. I'll be an *alien* in my own land.

But I have to leave this ship. Now that I'm awake, I must face whomever and whatever awaits me. I must bid farewell to this little pod—my friend who rescued me from the cold, cruel, grimy, sludgy hands of death on Vohaul's asteroid. Besides, the smell here is really getting to me. Obviously, air-quality control standards haven't changed in the past ten centuries. Oh, well. Good-bye, little pod. I must go. . . .

"Hey, people! I'm Roger Wilco!" If it weren't for me, you wouldn't even exist! Yes, I, Roger Wilco, single-handedly saved this entire planet and uni— "Huh?" Where is everybody? What the heck is *this* place? "Hellooooooooo . . . anybody heeeere?"

Where's the welcoming committee? I don't believe it. Of all the darn luck. Here I am, a universal hero, and not a soul is waiting to greet me. Boy, things have sure changed since I left. Judging by the looks—and smell—of this place, Xenon hasn't seen the likes of a janitor in . . . well, a thousand years! I can't believe it. I'm gone a couple of years, and the entire planet is in a shambles. Well, that pretty much proves it. I was the best darn janitor this place ever had.

Now, if I can just find my way around. I've never seen so much junk in my life, but why was it all dumped on Xenon, and where the heck is everybody? Oh, wait a minute; I'll bet I know. Those lazy, good-for-nothing. . . . They just figured since I used to be the janitor, I could clean up everybody's mess as soon as I woke up—and they've been letting it pile up for a *thousand years*?! Well, that's just great. We'll see about that. I'm not touching one thing. They can clean up their own mess. I quit!

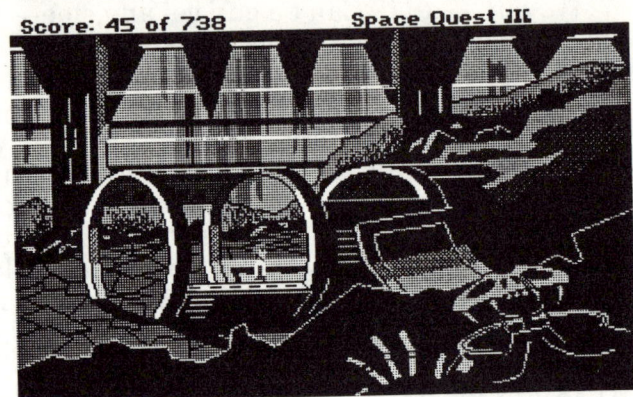

Score: 45 of 738   Space Quest III

So this doesn't look like the Xenon you remember? Perhaps that's because it isn't. And, yes, there is a lot of junk lying around, but just remember: One man's trash is another man's treasure.

What's that on the other side of the room? It looks like a tunnel of some sort. Hmmm, I don't remember seeing anything like that on Xenon or Orbital Station 4. Come to think of it, the layout of this place doesn't look familiar at all. And what's that rail thing overhead? That's a pretty odd place for a monorail.

Gee, maybe I'm not on Xenon—or on any of its orbital stations. Maybe I'm not even in the same galaxy! Then what is this place? All this junk—rotting garbage and burned-out fuselages? This place looks like a . . . a. . . . Oh, no! It can't be! Don't tell me I've been picked up by a . . . *garbage ship!* I'll bet that's what this is. Do I feel stupid. I'll bet a thousand years haven't passed. I'll bet a thousand *days* haven't even passed. Probably more like a thousand *minutes* since I fell asleep. I wouldn't be surprised if some of this crud came off Vohaul's asteroid.

Now I'm really in trouble. I might as well forget about finding a way off this tank. There's no way I'll find a usable spaceship on this dump. I'll be lucky if I can find anything that's even three-quarters intact. Hey, but maybe this isn't one of those sorry *robot* ships. Maybe this one is operated by people! . . . Nah, no way. Only a robot would be stupid enough to pull in a space pod carrying a live being. I'd better make sure I don't run into the one that's running this barge.

If I remember correctly, those trash freight operators are pro-grammed to sterilize anything alive—a form of rodent con-trol—unless it's wearing a security clearance badge. Great, Wilco. Once again, I've managed to go straight from the fry-ing pan into the fire. I don't *have* a security clearance badge; I was never responsible for garbage ship inspections.

Might as well get on with it anyway—I can't stay here forever. Let's see; I was planning to search through that tun-nel over there on the east side of the room when my mind started wandering. Guess I'll continue with that plan—what else is there to do?

Tunnel Passage

Well, nothing much in here to see—just a bunch of torn-up panels and some loose wire. Hmmm, this piece of wire looks to be in pretty good shape. Maybe I should grab a couple of strands—never know when it might come in handy. Besides, I've been pretty successful with the other junk I've collected during my travels. Who would have known a bathroom plunger would be the catalyst for my escaping that pool of acid on Vohaul's asteroid? And what about that piece of glass back on Kerona? If I hadn't had the brains to take it when I saw it, I never would have made it through that laser beam in the underground cavern. Yes, indeed, I believe I'll take whatever I can get my hands on in here. And wire—that's a pretty sensible thing. OK, now where to? I suppose I should see what's on the other end of this tunnel.

Robot-Head Room

Hmmm, now that's a strange-looking thing over there. It looks like some kind of giant robot head. Whooo, I'd hate to have been near that creep when it was alive. As a matter of fact, if I weren't as brave as I am, I could be pretty frightened by some of these robot skeletons I've seen around here. Gee, but what if some of them aren't r-robot skeletons? Wh-what if some of them are PEOPLE skeletons . . . *(gulp)* . . . I don't care to find out just this minute. I believe I'll pass on close in-spection of any skeletons from here on out.

OK, Wilco, once again, get a grip on yourself. I've only just begun. I can't start wimping out now. And I do know for a fact that that head over there did belong to a robot and not a person. It's so huge; I wonder if there's anything inside it. Well, there's only one way to find out. Besides, I don't see anywhere else to go in this area. It looks to be pretty much a dead end.

Score: 50 of 738   Space Quest III

You've managed to hold on to your sanity up to this point. However, the slightest little slip-up could send you flying over the edge.

Whoa, whoa, whoa—I almost fell! Apparently, this room does lead somewhere—straight down. I'll be more careful next time. There's so much old oil and gunk splattered on everything; I can't believe I haven't slipped until now. I'll just be very careful as I climb up the side of ol' metalface here . . . and take a peek inside the nonexistent eye. Hmmmm, this is interesting. It *does* lead somewhere else!

Trashed Battle Cruiser Room

Well, this is like climbing down a ladder. That was easy after all. I just have to remember to be careful. Hmmmm, what do we have here? Oh, brother. I knew it—some of this debris is from Vohaul's asteroid. Didn't I say his architect must have gone to Tinker Toy U?

Hey, there's a small pod-type vehicle near the west wall, and this thing here looks like a fairly recent model battle cruiser. Unfortunately, it looks like it *lost* the battle. I think I'll check out that smaller pod. It's probably much simpler to operate—since I never did become an engineer or anything.

The Pirates of Pestulon: St. Elmo's Ire □

Forget the giant Tinker Toys—this is no time to play. You're looking for a way out of this treasure trove, remember?

Gee whiz, I can't even get high enough to see inside this thing. This is ridiculous. All this junk lying around here, and I can't find anything light enough to move that's also big enough to stand on. Darn it. The glass is so dirty I'd need ammonia and about 20 Handi-Helpers to get rid of that grime. Shoot, I might as well look at that battle cruiser—or whatever it is.

I can see the pilot's seat through the front window. Hey, what's that in the seat? I can't get close enough to get a good look. Oh, it looks like something sticking out from under the cushion—some kind of copper or something. That really has me curious, but how do I get in? I can't even climb on top of the thing it's so covered in oil. Of all the darn luck. You'd think there would be at least an old ladder or portable steps or something around here I could use, but noooo, there's nothing. This floating heap of garbage is the sorriest trash dump I've ever rummaged through. I can't do anything in here.

I've looked all over the place, and it's the same old story—nothing. Might as well climb back up and go back to my own space pod. Maybe I can find something to use in that pile of junk I saw to the south of my pod. First, I'll have to climb back up this wall through the robot's eye, then back through the banged-up tunnel, past my poor little pod, and on to the south end of that room. . . .

Trash Pile Room

Well, I see that big pile of trash didn't go anywhere—ha. What did you think, Wilco, it would sprout legs and walk away? Oh, you're not funny. Geez, I can't even make myself laugh. I wish I *could* see some humor in this situation. I guess it's pretty ironic if I think about it, ending up like this, stranded on a garbage dump. It's as though the fates are against me. I guess I'll be a janitor forever.

Let's see; maybe there *is* something in this pile. . . . Eewww, yuck. The smell gets even worse down here. What do they process on this tank, sewage? This reminds me of Vohaul's asteroid. Well, I don't see one useful thing in this entire heap of garbage. . . . Saaaay, there's the *Jupiter 2*. Wow, I can't believe it. You mean to tell me after all those years of being lost in space, the Robinson family ended up here? Gee, that doesn't exactly say much for the *Jupiter 2*. Maybe they traded it in for a better mod. . . . Hey, what's that noise coming from over there? I'd better check it out.

Buckets / Conveyor Belt Room

Wow, look at those buckets! I wonder what that's all about. I'm about to find out; that's for sure. I actually feel like I'm getting somewhere now. I'm sick and tired of running into dead ends.

Hmmmm, this contraption brings up buckets of trash from somewhere—heaven only knows what *that* place smells like—and then loads them onto that conveyor belt up there. It looks like the trash is then dumped into some sort of shredder or compactor or something. Wow, what a neat operation. I wonder if the stuff is recycled. I could have used something like this back on Orbital Station 4. Wait a minute; if they're so efficient here, why the heck is all this other stuff piled up against the walls and lying all over the floor?

What difference does it make, anyway? I don't plan to hang around this joint any longer than I have to. There must be a way off this ship. The conveyor belt—it runs so close to that rail. Hmmmm, I'll bet I could jump from the belt to that rail. It would be close—those piles of trash don't ride for very long before they're dumped into that shredder—but I'll bet I could do it. At least it would give me a chance to see the layout of this place from overhead. That way, I could plan my escape a little more easily.

OK, here goes. I'll just step into the next bucket that comes up . . . and . . . here I go. Hey, this is kind of fun! But I'd better brace myself for a good jump as soon as I hit that belt! Almost there . . . almost there . . . I'm there! Y-y-y-YIKES! This thing is shorter and faster than I thought! . . . I have to stand up! . . . Whoa, whoa, whoa . . . ahhhhh! JUMP! . . . Holy chopped liver! That thing almost canned me—literally!

Maybe I was stupid to try that jump, but I didn't have a choice. There's no other way to get a good look at this place. I have to find an escape route. Maybe I'll just sit here a minute and catch my breath before moving on. Gosh, I can't quit thinking about what almost happened to me. I've come closer to death than that before, so why can't I get the picture of a canned Roger Wilco out of my brain? Maybe it's how I would have died. I can see it now on the front page of the *Xenon Times:* "Missing Janitor Found 'Ex-Sponged' on Trash Freighter." And the radio stations, they'd go on and on and on. I can just hear it:

> *The compacted body of Roger Wilco, the Xenon janitor reported missing over a year ago, was found today aboard an orbiting trash freighter. The robot barge was making its weekly garbage pickup at Xenon Orbital Station 4, Wilco's assigned employment station at the time of his disappearance. The body was found among a stack of compacted garbage blocks during a routine inspection of the freighter. . . .*
> *Those wishing to pay homage to the late janitor may donate cleaning fluids to the Sanitarian Association of Greater Xenon.*

Yeah, that's how they'd remember me—donating cleaning fluids. I know how that works. They wouldn't even remember the golden mop I received or the ticker-tape parade.

On the Overhead Rail

Now, which way should I go? Heyyy, this doesn't give any great view up here. It's just the same old stuff—piles of trash—only I'm looking at it from a bird's eye. You mean I risked my life for nothing? Great, that's just great. Oh, well. I'm stuck here now; might as well find a way down. Besides, maybe I'll get lucky just this once and find an opening that isn't visible from the ground.

Not only are you no James Bond and no engineer, but you're no tightrope walker either. Save the acrobatics for when your feet are a little closer to the ground.

It looks like this thing circles all the way around, so either way, I'm going to end up in the same place. Might as well take the shortest route—to the west. This is going to be a difficult enough balancing act without forcing myself to turn corners. I'll take it slow and easy until I reach that doorway-looking place over there. I can't tell what that is—must be some kind of room.

Hey, maybe it's a control room. . . . Yikes, what if it isn't empty? Once I reach that doorway, I'll just take a peek around the corner before doing anything else. OK, Wilco, you're almost there. Don't blow it now by falling off. . . . Saaaay, what's that little chair thing hanging under the rail? It must be the rail transport! Duh, no kidding, Wilco; what else would it be? Of course, it's some kind of rail transport—I just wonder what it does. What the heck? It's my chance to get a good view of this place. It might even be fun taking a ride on that little chair.

Uh-oh. What's that over there? I knew it. Drat. You can walk miles and miles on any planet, spaceship—or TRASH

freighter, in this case—and never ever see another living being. But you can bet your sweet bippy there's going to be
some clanging bucket of bolts running around just waiting to
annoy the heck out of you and ruin everything. Of all the
darn luck. Who is he, anyway—the major-domo? Ha—more
like a major headache. Now how am I going to get on that rail
chair before that nincompoop spots me and decides to STER-
IL-LIZE. Sterilize schmerilize. Hmmmp.

Hey, what's that? It looks like some sort of chute—but
leading to where, I wonder. Maybe I'll just take a chance and
tiptoe over there to get a closer look. Ol' radarhead looks
pretty engrossed in what he's doing. Hmmm, that must be
some sort of garbage chute. Of course, it's a garbage chute.
What other kind of chute would be aboard a *garbage*
freighter?

Uh-oh. I have a feeling I've been standing out in the
open just a little too long now. Winking-blinking-and-blip
looks like he's stopped . . . pushing . . . buttons. . . . Well, it
looks like I've got to jump. I have no choice. It's down the
chute I goooo!

The Rat's Lair

Oh, no. Wh-what k-kind of spooky place is this? And I mean
spooky. What are all these wires and things hanging around
here? I can hardly see. I'll bet this is some type of torture
chamber or something—or where they dump all the "accidental" death victims. One thing's for sure; it *smells* like an
accidental death in here.

Huh. Wh-what was that noise? Something moved. I saw
something move up there in the rafters! . . . There it is
again! . . . There it is *again*. Wh-what could it be? This place
gives me the willies. The next thing you know I'll hear
chains rattling and see g-g-g-GHOSTS! My heart sounds like
a bass drum in my ears.

Huh! What was *that*? I s-saw something move up there
again! It's watching me—whatever *it* is. No matter what I
thought before, now I know for sure—I've got to get the heck
off this freighter—or at least out of this part of it. I need to be
brave now and start looking around for a way out. I see a ladder over there. I sure could use that to climb into that old
wrecked battle cruiser, but I'm too scared to g-go over there

and get it. Th-those wires overhead—maybe they lead to an opening somewhere. I'll just f-follow this wire and see what h-happens.

Oh, please lead to a way out of here! . . . Ouch! Oh, great. This wire leads straight to a wall—the west wall. Well, that's a fine kettle of fish. What am I supposed to do now, create my own doorway? Hey, what's this? An *opening* in the wall—or more like a hole, actually. So that wire leads somewhere after all.

Oh, gee. In my excitement over hitting my head and finding this hole, I forgot about being scared. Now I remember. OK, Wilco, just try to concentrate on getting out of here, and maybe you'll forget it again. Think positively . . . positively. . . . Maybe I can convince myself that I shouldn't be afraid and that this isn't really a torture chamber or a dead-body repository after all. And dare I stick my hand in that hole?

A hole in the wall. . . . Gee, this reminds me of that hole in the cliff on Kerona. I never did get the chance to go back and see what was in it. Well, I passed on that chance, so now maybe I should risk facing the unknown by sticking my hand in there and feeling around. I have to do this. I *have* to. How do I know my life doesn't depend on this one little hole in the wall? O-K-K-K, here g-g-goes . . . but so help me, if there's a scorpazoid or a spider in there, I'm going to scream bloody murder!

Ahhhhhh! . . . What happened to the lights? Oh, uh . . . ahem. *This* is what happened to the lights. That little hole in the wall houses the little reactor that generates the little light in here. And now I'm holding the reactor because I just pulled it out of the wall. So no spiders and no scorpazoids in *that* wall, thank goodness. I'd better hang on to this jewel—it might be useful if I can ever find a way into that cruiser.

Now to find my way out so I can get *back* to the cruiser, which certainly won't happen by my standing around here. I have to be brave. I'm sure I can make my way over to that ladder—even though it's black as night in here. OK . . . I'll take it step by . . . oops! . . . step . . . ouch! . . . Darn trash on the floor. Hope I didn't step on any bones . . . BONES? Maybe I'm walking all over the Robinson family's bones!

Now just hang on, Wilco; you're almost there. There couldn't possibly be any bones on the floor. I'm getting myself all worked up over nothing. . . . The ladder! I found it! Hey, there's an opening at the top of the wall! All I have to do is climb up this ladder, and I'm outta here!

Out of the Rat's Lair

Well, gee. No, I don't feel foolish at all. Just because there were no scorpazoids or spiders in that hole. Just because I got all worked up about being in a torture chamber and walking on dead bodies, even though I'm on a garbage freighter where there is no such thing. That's OK. No problem. Yes, my sanity left me for just a minute—went south of the border—but hey, I'm allowed to feel a chill up my spine every now and then. What's the crime in that? I'd like to see somebody else—*anybody* else—go through what I've been through the past two years and see if *they* can keep from having one off moment.

Well, what do you know? I'm back at the old trash pile again. Let's see if I can remember which way to go to get back to that battle cruiser. Hmmm, I believe I go north from here—yeah, that's it—and then through that trashed tanker's tunnel, and then I'll have to crawl through the broken robot eye again. Shoot, why can't there be a shortcut or something? I can't believe I have to walk all the way back there again, not to mention risk falling off the edge of that giant robot head.

Oh, well. Haste makes waste, and I don't need any more waste—I'm already wallowing in it. . . . OK, here we go . . . ooops! I almost forgot the ladder. Boy, that'd be brilliant, Wilco. No ladder, no getting inside the cruiser! Gee, it's kind of heavy—clunky, too. How am I supposed to cart this thing around? Well, whaddaya know? This is some type of portable ladder! Let's see—fold it, bend it, twist it, fold it again, twist it again, and then snap it into place. Oh, brother—Ronko city. Oh, well. I certainly can't and won't complain. I'm lucky to even *have* a ladder. I'll have to give credit where credit is due. Stuffing a compact ladder into my pocket is a lot niftier—and lighter—than hauling one around on my back, especially while trying to make it through some of the trickier passageways around here.

Back in the Tunnel Passage

OK, time to hit the road, Jack. . . . Yep, this is the way, all right. . . . There's my escape pod—what's left of it. Catch you later, dude. . . . And now back through the tanker tunnel . . . past the same old rusted wires hanging from the—"HEY! Ouch! Ooooh! Hey, what the— OUCH! . . . Hey, you! You dirty rat, you! Come back here with my stuff! You lousy RO-DENT! Hey, your mama's a lab mouse!"

Score: 50 of 738   Space Quest III

Oops! And you thought that guy from Ulence Flats had no honor. That's what you get for stealing from a thief.

Jerk. I don't believe it. That dirty, rotten scoundrel—he took my reactor! So that's what I heard moving around in that room. I should have known—a bunch of lousy pack rats rummaging around. And to think I was afraid—torture chamber, indeed. Now, darn it, I've got to retrace my foot-steps and go *all* the way back to those trash buckets, and then get on that conveyor belt, and then jump up on the rail, and . . . whew, I'm tired just *thinking* about it. I'm starting to feel like a yo-yo with all this going back and forth.

Well, I don't have a choice. I have to get that reactor back. Dollars to doughnuts it's in that same hole in the wall where I found it. . . . Whoa, wait a minute. What am I talking about? I still have my ladder—and I know how to get to the rat's lair through that trash pile. Hot dog, I don't have to go back to the conveyor belt after all!

OK, I know that opening's around here somewhere. It can't be that hard to fiiiiii . . . whooops! I think I found it! . . . This is it, all right. OK, now to unfold my handy-dandy lad-der and place it down in the opening. Now, I'll just climb down and pray that none of those sorry wingless bats try to ambush me again. I won't have it. I just won't have it!

Back in the Rat's Lair

Well, one thing's for sure; the lights in this room didn't turn their . . . I mean themselves back on, which can only mean one thing—the reactor's back in the hole. Gee, Wilco, thanks for that blinding glimpse of the obvious. OK, chump, you want to play this little game of tug of war? We'll see who wins. I've beat much greater odds than you in my lifetime— at least over the past couple of years I have.

Just as I thought, here's the reactor—and there go the lights again. Well, that about wraps this up. Oh, wait; maybe I should check to see if anything else is in there. No telling what sort of odds and ends are lying around this place, considering the purveyors. . . . Heyyyyy, what's this? My *wire!* "You thief! I don't believe it! You good-for-nothing. . . . *You* ought to be running this ship, seeing as how you're so good at collecting garbage! And I know you hear me—I can practically see your beady little eyes glowing in the dark." Boy, I'm glad I thought to check that hole again. Who knows when I might need that wire! Now to get out of this rathole for the second time—and I'd better remember to bring my ladder, too.

Out of the Rat's Lair—Again!

Now how did I fold this thing up the last time? . . . Hmmm, let's see . . . shoot! It's not working. This is as bad as trying to refold a map. Let's see. . . . Turn it, twist it, fold it. . . . That's it! . . . And snap it into place, thank goodness! That ladder's a little bit of a pain—but worth it. I'm not complaining. Things could be much worse. Now back to the battle cruiser.

Oh, no. A thought just occurred to me—what if that darn rat tries to ambush me again? Hmmp. This time I'll know what to expect, and *this time* he's not getting away with it. I can stand up for myself. I'll just tear that little monster limb from limb. This time, he doesn't know what kind of ambush *he's* walking into! . . . I'm almost to the tunnel again. OK, rat-fink, you'd better be prepared for the battle of your life. . . .

Hmmm, that's strange. I'm almost through the tunnel, and there's no sign of any rat. No sign of anything period.

☐ Chapter 7

Well, I guess he took one look at me and decided not to risk it again. Must have seen how I was pretty burned up over that little charade. I'm sure he saw the fire in my eyes and decided it wasn't worth it. I might be a janitor, but that doesn't mean I can't fight a good battle.

Inside the Battle Cruiser

Okey-dokey, I'm back. Now, to unfold this ladder and set it up against the ship so I can climb in. Geez, this cruiser's covered in grease—yuck. I'd better be careful—wouldn't want to fall and break my neck!

OK, I'm on top of the ship. Now I'll just pull . . . the latch . . . up. It's open! Just a few short hours ago I didn't think I had a prayer. Now I'm standing inside this nifty battle cruiser. I'll just keep my fingers crossed and hope this baby's in working order.

Let's see; what do we have in here anyway? I see a couple of passenger seats—either that or this cruiser was built for a crew of three. Over there, of course, is the pilot's seat I saw through the window. And what's this—a computer? Hmmm, that may be helpful at some point. Well, everything looks pretty good to me—the interior seems to have survived a few battles. That's a very good sign—I just hope the engine's in as good a shape as the upholstery. OK, pilot's seat, here I come.

Now that I'm seated, I should try the engine . . . so here goes! . . . Great, nothing. I don't believe it. On second thought, I *do* believe it. This is par for the course. Now what do I do? Hmmm, maybe the ship's computer is still alive. Darn it, why isn't that computer up here near the pilot's seat? Now I have to get up and go all the way to the back of the ship. What kind of lousy design is this anyway?

OK, computer, do your stuff. Please, please, please tell me what's wrong. . . . No wonder this ship is grounded—there's no power supply. Now isn't that just peachy? I should have expected something like this. Every time I think I'm getting somewhere, my hopes are demolished by another turn of bad luck. In fact, I'm beginning to believe there's no such thing as good luck—not for me, anyway . . . (sniff).

I'm just another Wednesday's child . . . full of woe . . . (sniff) . . . what's the use in . . . (sniff, sniff) . . . fighting it? I have to live the life I was born to live—be it a leader, a hero, or a . . . *(sniff)* . . . sad-sack janitor. All I have to my name is a . . . a little glowing gemstone, a useless piece of wire, a . . . a reactor for a couple of dim, dusty, light bulbs swinging from rusty chains on the ceiling of a . . . a . . . REACTOR? A reactor! I have in my possession a tiny little reactor that's bound to supply ten million times its weight in power!

Dollars to doughnuts that lousy pack rat took it from this ship to start with. The winds of fortune may finally be blowing in a new direction for poor old Roger Wilco! OK, this computer tells exactly how to install a tiny *reactor* that will supply power directly to the engines. Great. . . . OK, I've put it in place, but the cable's too short, darn it. Don't tell me I've come this far . . . oh, wait a minute—with all the trashed space vehicles piled up on this barge, I'm bound to find a substitute for a cable somewhere. All I need is something with a—*wire!* Oh, Wilco, you clever devil, you. I already have a wire!

Gee, I hope this piece of wire is long enough. . . . It fits! I'm in business again! Back to the pilot's seat to fire up these engines, and we'll be going . . . nowhere. Oh, no. I should have known. What is it this time? Another darn problem. I should have known there would be more to this than meets the eye. What could it possibly be? I'll have to check the computer again, darn it.

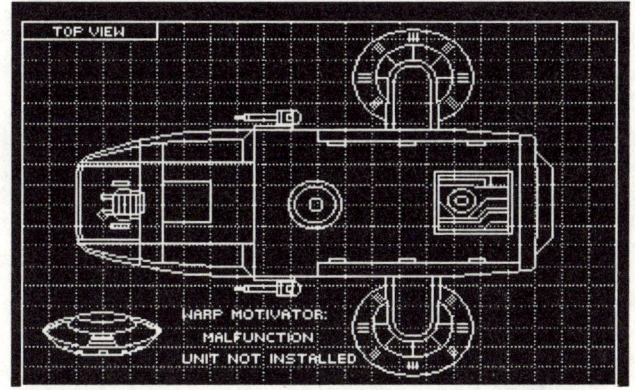

This should "motivate" you to do a little more rummaging around.

OK, what now? Hmmm, the warp *what* is missing? Warp *motivator?* What the heck is a warp motivator? Maybe it motivates the ship to warp power. Duh, thank you, Wilco, for another blinding glimpse of the obvious. Well, I suppose it

would be nice to have warp power—if I plan on going anywhere. Let's see; it shows some saucer-looking thing in this
diagram. Well, that should be easy to find. Matter of fact, I
think I remember seeing something like that back in the area
where my escape pod was dumped.

Fine. This should be easy as pie. I'll just *motivate* myself
back over to my pod, pick up the saucer thing, bring it back
here, and see if it fits. Nothing to it—and hopefully, this will
be the *last* time I have to make that journey again.

Oops, I'd better be careful getting out of this ship—I almost forgot how greasy it is on the exterior. I'll take it
veeeery carefully climbing out and down the ladder. Hmmm,
I wonder if I'll need that ladder again. . . . Nah, I should leave
it here. This is the only place I've needed it so far—except in
that rathole. Oh, on second thought, I'd better take it with
me. "I wouldn't want a certain *RODENT* around here to steal
my ladder or anything."

Back in the Trashed Pod Room

I'm *really* getting tired of this going back and forth, back and
forth. Whatever happened to simplified floor plans, or better
yet—organized garbage piles? Saaay, there it is—that saucer-
looking thing that I'm sure is the warp motivator.

OK, little fellow, I'm going to take you . . . mmmmff
. . . ooh . . . OOOOH! Rats, I'm going to break my back! I can't
lift this thing! It must weigh a ton. Of all the sorry. . . . Well,
how the heck did it get over here, anyway? No way that ro-
dent could have moved it from the cruiser. Well, that's a
mystery for sure. Hmmm, now what am I going to do? I know
one thing; I'm not about to give up after coming this far. I'm
so close to being out of this can I can taste it.

There must be a way to move it. Maybe I could find a
rope to tie to it and then somehow throw the rope over those
rails overhead, and . . . saaaay, wait a minute! The rail chair!
That's it! Why didn't I think of it before? That chair isn't for
aerial views of trash piles—it's a brand new, shiny Galaxia
U-Do-It Crap Grabber! Boy, I remember reading about those
things in *Janitor Today*, only then, the technology was still in
its beginning stages. Wow, maybe I was asleep longer than I
thought! These babies were supposed to revolutionize the
entire trash freighter industry. . . . Of course, with a bliphead

robot at the helm, I can see why it hasn't made any difference aboard this ship.

I'll just make a beeline back to those buckets, take my chances on that conveyor belt again, and hitch a ride aboard the crap grabber. I know where to find the warp motivator—I just hope I can do this quickly without being detected by you-know-who. At least coming back to this room wasn't a wasted trip—now I'm only a hop and a skip away from the conveyor belt room.

Back at the Buckets

Oh, boy. I didn't realize how scary this would be the second time around. I don't know if I can do this again. I almost had my insides turned out last time. Gee, this is worse than getting on the double Ferris wheel at World O' Wonders.

O-K-K, here goes *nothing!* . . . Just take a deep breath, Wilco. I can do it this time—I did it once before. I just have to remember to stand up as quickly as possible when I hit that conveyor belt and then jump! I'm almost there. . . . OK, this is it . . . JUMP! I did it! I made it again! I sure am glad it doesn't take practice to make perfect—just a couple of seconds more, and I'd be . . . *(gulp)* . . . history!

On the Rail Again

Oh, well. That's over. I'm safe and alive—well, alive anyway. I don't know about safe. This is kind of like a high-wire act. I just have to be careful once I get to the grabber— wouldn't want to fall while trying to get in. I can't let that robot see me, either. I'll just tiptoe as quietly as I can—there he is, but he still doesn't see me—and make my way into this little vehicle here. . . . I'm in! Good!

Let's see—just as I thought—this is indeed a Galaxia Crap Grabber. There's the claw button. I didn't recognize this vehicle before because the claw is hidden until lowered. What a nifty little setup. Now, which way should I go? I can see the warp motivator from here, but obviously I'm going to have to take the grabber all the way around to the other side. Hmmm, should I risk taking the shortcut around the robot's quarters? Oh, heck. Why not? I'm sure I'll be moving too fast for that sucker to see me.

And away we go. . . . Gee, this thing makes enough noise. . . . Whew, I made it past that walking radar dish . . . and I believe I can reach the warp motivator from just . . . about . . . here! Great, I'm suspended directly over it. Now to lower the grabber claw. . . . Got it! I got the warp motivator! I got it, robot scum!

Full speed ahead now—I can't get this thing installed fast enough, as far as I'm concerned. Yesterday wouldn't have been soon enough for me. OK, this may be a little tricky going around these curves. The ship's just below that second turn. . . . Better slow it down now . . . and . . . stop! I just hope this works. First I have to lower the claw—there she goes. Wow, it looks like I picked the right piece of junk to install— it's a perfect fit!

Now what? I have to get out of this grabber and back down to the battle cruiser. Shoot, looks like I'm going to have to jump down that chute again. Maybe the head honcho is tending to other duties by now—at least I hope so. I can't help but wonder when my luck's going to run out as far as that robot's concerned. I've snuck by his quarters one too many times, it seems. I'm sure the odds are catching up with me, as my luck goes. Oh, well. I have no choice. But wait—I can make this quick! I'll just park this grabber in front of the chute and jump! . . . Well, there's the robot . . . and there's the garbage chute . . . and here I gooooooooooo!

Back in the Rat's Lair—Again!

Oh, geez. I forgot about ending up in this wasteland. I wonder what my sticky-fingered friends are up to now. Well, at least I know my way out of here. "I certainly should know my way out by now, considering HOW MANY TIMES I'VE HAD TO COME IN HERE! Hey, you guys! May all your paths lead to one giant maze!"

And now, to unfold my handy-dandy . . . whoa, I just had a thought. I almost didn't bring my ladder with me when I went to get the warp motivator! I didn't realize it would be too heavy to carry and I'd end up in here again. If I hadn't brought it with me, I'd be stuck in this rathole right now— forever—with no means of escape. I would never have remembered to go back and get the ladder before climbing back on that conveyor belt.

Oh, my gosh. It makes me faint of heart just to *think* about it. My fate would have been sealed right then and there—by my own hands. I would have died in this place. There'd be nothing left of me here but . . . but . . . *bones!* And they'd be on the floor, mixed in with all this junk. And someday, someone would come along, just like I did, and *step* on them!

Wh-what if that actually did happen to someone in here, just like I thought before. What if it happened to the *Robinsons?!* I knew that idea wasn't too farfetched. I could be walking on Will Robinson's bones right now! Y-y-y-yikes! Get me outta here!

Out of the Rat's Lair—for Good!

Boy, that was close. If you ask me, I was right all along. I'll bet a million buckazoids there were bones all over the floor in that . . . that . . . *place*. Thank goodness I'm out of there now, and hopefully, I'll never have to return. I still can't believe I *almost* left that ladder back at the spaceship. Better fold it up and put it in my pocket right now before I forget. This ladder has saved me more than once now. It's my *only* means of getting inside the battle cruiser—speaking of which, I'd better get back there ASAP.

Now that I'm so close to leaving this place—at least I *think* I am—I don't want to risk anything else that could prevent me from leaving. Oh, gee. I sure hope that rat isn't planning any more surprise visits. That's the last thing I need.

Back Inside the Battle Cruiser

Well, bless my soul. I made it back without anything else happening. No rats were in the tunnel, I didn't fall off the edge while climbing through the robot eye, and I didn't slip while climbing into the cruiser. It's about time something went right for a change.

Fortunately, by the look of things, I've had another stroke of good luck—not only does the cruiser seem to be in working order, I found a few buckazoids stuffed under the cushion of my seat! Monolith Burger, here I come!

Let's see; now I need to start the engines, and wow, this really is a neat ship! Golly, I've never flown a battle cruiser before. Heck, I've never flown much of anything but idiot-proof escape pods and that sorry excuse for a shuttle that plunged me into the pits of sludge.

Hmmm, I might as well use whatever controls I can—the radar for one. I'll need to track alien warships—any that dare to cross my path. I'll zero in on them using my radar screen and then ZAP, ZAP, ZAP! BANG, BANG, BANG! And they'll be gone for good!

You haven't heard the rhapsodic strains of any fat ladies recently, have you? Then don't be too quick to kick off your shoes just because you cleared the freighter.

Let's see; what else can I do here? How about Take Off— that's always a good one! Oh, boy! The engines are fired. . . . I'm actually moving! The ship is lifting off the floor of the tanker . . . I've cleared the floor! I'm on my way! OK, baby, let's go! Let's get this show on the road!

Heyyy, wait a minute. We've stopped—I mean, *I've* stopped—I mean the ship's stopped. I don't get it. I'm hovering in midair. What's the matter with this thing? Oh, come on, go! Go! . . . Darn it! What's wrong with it now? Everything looks OK—at least it does to me, no engineer that I am. . . . Oh, wait a minute; *that's* it! I've got it! No wonder we're not going anywhere. I forgot—there's a ceiling overhead. Wilco, you beanhead, you're still *inside* the garbage freighter.

Well, I don't know what else to do. I'll just have to blast my way out of this tank. Oh, boy. I really get to use the weapons system! I've always wanted to pilot a battle cruiser! This is great! Let's see; how would Captain Kirk handle this?

"Battle stations, battle stations. This is not a drill. Repeat, this is not a drill . . . front shields . . . prepare to fire photon torpedoes . . . fire when ready . . . FIRE! It's a hit! We got him! Prepare to leave garbage freighter . . . warp factor 3

. . . that's it! We've cleared the garbage freighter! Cancel red alert. Secure to general quarters. . . ." Wow! I did it! I'm out of there! Sound the trumpets! Roll the drums! Roger Wilco is back!

Whew, what a relief. All that time—I can't believe I'm finally out of that pit. Well, I guess I can cancel the front shields—now that I've canceled the red alert.

Now where should I go? I suppose I should check the radar to see what's out there . . . hmmm, this is pretty self-explanatory. I should press Scan. Yes, that's it. . . . Wow, this is even neater than I thought. Look at that radar! Hey, what's that? The radar has zeroed in on a planet called *Phleebhut.* I don't remember any planet by that name in Xenon's solar system—or even in our galaxy for that matter. I guess it's possible I'm in a different galaxy, but I find that awfully hard to believe. Maybe Phleebhut's been around all along, and I just haven't heard of it.

Gee, it must be some really hopping place if it's so obscure that I've never even heard the name. Oh, well. I'll just set the course to Phleebhut for now. I might as well see what's there. Oh, boy! I get to use light speed, too! Here goes. . . .

Those jerks back on Orbital Station 4 would never believe this, that's for sure. . . . Wha . . . huh? Wh-wh-what was that? Some kind of alien spaceship j-just flew by, and I could see something f-flashing on its screen. It looked like the word *terminate*—and he was l-looking at me . . . *(gulp)!* Oh, no. What did *I* do? I haven't d-done anything to anyone! What would someone like that w-want with me?

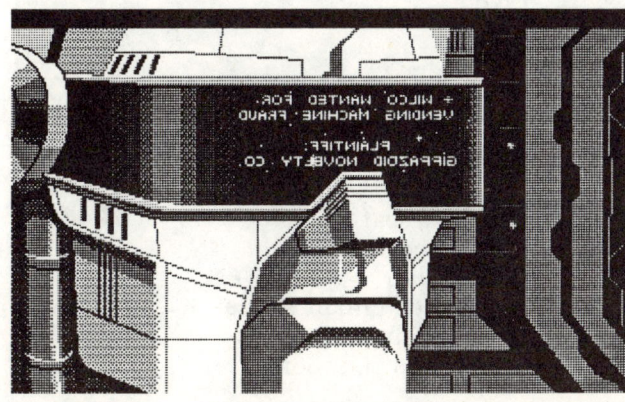

It says, "WILCO WANTED FOR VENDING MACHINE FRAUD. PLAINTIFF: GIPPAZOID NOVELTY CO." Uh oh, this could be terminal!

141

Maybe that message—whatever it said—wasn't meant for me. That's it. There's no way it could be for me. Heck, I've been stuck on a trash freighter for who knows how long. There's no way anyone could know the exact time or even *if* I had escaped. I'm getting all worked up for nothing. That alien made a mistake—obviously mistook me for someone else—someone I sure wouldn't want to be.

No, I won't worry anymore. I'll just rest until I get to Phleebhut. Maybe I'll actually find some civilization there. I sure hope so. A fellow can get mighty lonely wandering aimlessly through space. . . .

Well, it looks like I've arrived. Time to land this baby . . . and see what bad luck I'll have in *this* place.

Game Points Earned

Action	Points
Blasting out of the trash freighter	25
Climbing through the robot's eye	5
Entering the battle cruiser	10
Finding and taking the reactor	15
Finding buckazoids on the cruiser	10
Getting the ladder	10
Getting the wire	5
Grabbing the motivator	15
Installing the motivator	15
Installing the power reactor	5
Jumping to the rail	10
Riding in the trash bucket	5
Sliding down the garbage chute	5
Using the ladder at the battle cruiser	5
Using the wire	5

Object Locations

Object	Location
Buckazoids	Pilot's seat
Gemstone	Your pocket
Ladder	Rat's lair
Old wire	Trashed tanker tunnel
Power reactor	Rat's lair
Warp motivator	Trashed pod room

The Planet Phleebhut

Mission: Acquire more buckazoids, collect a few souvenirs, and annihilate the Annihilator before leaving.
Total Points: 103

The Planet Phleebhut

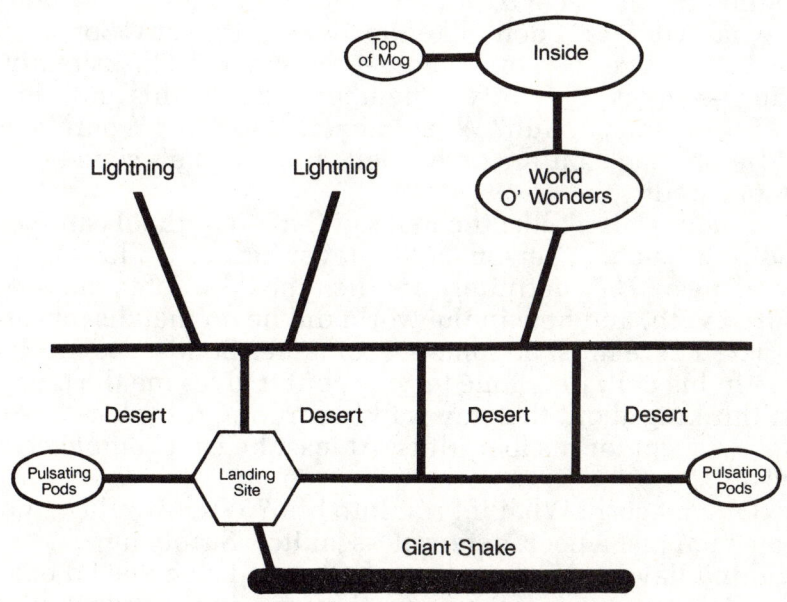

Inventory

Astro Chicken hat
buckazoids
invisibility belt
Orat on a Stick
Thermoweave underwear

At the Landing Site

So this is Phleebhut. For some reason that name is starting to ring a bell. It's an odd name for a planet—but I just *know* I've heard it before, now that I think about it. Oh, well. That must have been years ago since I don't remember it now. Maybe it will come to me later. In the meantime, for a place that's supposed to have at least one settlement, I don't see much in the way of civilization or anything else here.

Hey, another ship's landing over there. So there must be some attraction to this place after all. Maybe whoever it is can tell me . . . eeeeeyikes! Who the devil is *that* . . . it . . . he . . . whatever it is? I don't like the looks of that guy. Somehow I think I've seen him before—but where? This is really strange—a planet I faintly remember and now this android-looking person . . . huh? What happened to him? Where'd he go? He pressed a button on his belt, and now he's gone—disappeared!

G-golly, I don't like the looks of that—not that I can see anything. That . . . person, or whatever he was . . . looked pretty mean. He's definitely not the type I'd want to mess around with, and how in the world did he do that disappearing act? There must be some sort of James Bond–type mechanism in his belt. He's gone for now, but it gives me the creeps just thinking about that guy walking around this place unseen—except for his footprints. At least he isn't *completely* invisible!

Oh, my gosh! What if I run into him? What would he do to me? I'm just a poor, defenseless janitor. Surely he wouldn't have any reason to . . . (gulp) . . . harm *me!* I'd better watch my step just the same, though. Since he went off to the east, I think I'll take a hike to the west. Hopefully, civilization will be just around the corner.

West of the Landing Site

Hmmm, this is an unusual-looking rock formation. It's almost like a cave, only hollowed out. Eeewww yuck, what are those things clinging to the ceiling? They look like pulsating pizzas—or fake vomit or something. Now where have I seen something like that before? . . . Oh, *now* I remember. Oh, my gosh. No way, José, I'm going near those things. If they're the same kind of pizza monsters I've seen before, I know exactly what kind of damage they can do! I don't want one of them clinging to *my* back! The best thing for me to do is to get the heck out of here—pronto!

That was close. At least I know what sort of creatures are hanging around Phleebhut—no pun intended. Well, now that I'm back at my ship, I'll have to decide on another path. Looks like a storm is brewing to the north. It's *bad!* It looks real bad! No, sir, I'm not gonna go in that direction—not gonna do it. So it's either east or south, and I think I'll try south just in case the storm decides to move any closer.

Not only can lightning strike twice, but it can strike three, four, and five times if that's what it takes to get your compass adjusted.

South of the Landing Site

OK, there's a lot of desert so far, but no . . . wh-wh-wh-what is THAT? A tremendously huge, giant *snake!* That thing looks like it's about to make a meal out of me! Well, no thank you—I'm not about to become your or anyone else's dinner! Sorry, pal, I gave at the office. If you want a blood donor, you'll have to look elsewhere. Besides, you overgrown garden hose, I only give pints! Try picking on someone your own size next time . . . y-y-y-yikes! . . . See ya!

Another close call. How was I supposed to know I'd run into some gargantuan reptile? Darn it. The longer I'm here, the less I like the looks and feel of this place. So far I've seen some metallic chap do a disappearing act, a bunch of throbbing pizzas clinging to a rock, and now this mile-long snake that could swallow Orbital Station 4 in one gulp. I'd have to be an idiot to stay around here any longer.

The smartest move I could make right now is to get back in that ship and pretend like I've never even heard of

So "throbbing pizzas" aren't your cup of tea. Smart thinking, since they can be quite terminal. Speaking of terminal, isn't there someone else you could show them to?

Phleebhut . . . *Phleebhut* . . . *Phleebhut* . . . darn it, but I have heard of it. Why can't I remember? Oh, what the heck. This isn't the sort of place I plan to visit too often anyway, so what difference does it make? . . . Uh-oh, are those footprints leading into my ship? Y-y-y-yikes! That . . . *thing* . . . must be after my ship! What if he steals it? What if he isn't after my ship? What if he's after ME? I'm getting out of here on the double!

I don't have a choice; I have to go east. Maybe I can give him the slip—make him think I'm out in the desert somewhere. Then, when he isn't looking, I'll sneak back to my ship and . . . and . . . why the heck am I wasting time thinking about this? I've got to get going—now!

East of the Landing Site

Another close call! I've had a lot of those lately. I can't believe my awful luck. I thought things were finally looking up when I managed to escape that trash freighter, and now here I am again, faced with all sorts of unforeseeable predicaments—and dangerous ones at that! Well, now that I'm out in the desert, at least I don't see any more footprints!

Huh? . . . What was that? I saw something move behind that rock over there! Oh, no. Don't tell me. It can't be. They're even worse than spiders—well, almost. The point is, they're horrible! Please don't be a . . . a . . . ahhhhh! IT IS! It's a SCORPAZOID! One of the deadliest arachnazoids in the universe—and it's after me! Get away from me . . . I'll run south . . . no, east . . . no, north! I'll run any way I can to get away from that creepy-crawling thing!

In the Desert, Northeast

Dang, this place is crawling with natural disasters—and they're all alive and well—and after me! At least I made it out of those rocks and got away from that scorpazoid. I'll bet there are hundreds of them out in the desert. Speaking of desert, which way should I go now? Those must have been isolated electrical storms I saw to the north, since everything looks pretty calm over here. I'm definitely northeast of my spaceship now, and there's no sign of lightning . . . huh? Wh-wh-wh-what's *that*?

Oh, no. Don't tell me it's another reptile—only this time, it's a mile *high!* That thing could crush me with a single look! It's moving, too, and I'll bet it's headed this way! Oh, woe is me . . . (*sniff*) . . . what did I ever do to deserve all this? . . . (*sniff, sniff*) . . . I've tried to be a good janitor—mind my own business, stay out of trouble. So I snuck in a few z's every now and then. What's the crime in needing an occasional rest? . . . (*sniff*) . . . I might as well give up. This time I don't stand a chance. There's nowhere for me to go. That giant lizard, or dinosaur, or whatever it is, is still moving. . . . It's just waiting for me to run—they always like the chase, those big brutes.

Well, I'm not going to give it a chase! I'm not going to let it have any fun at all. "You hear me, you big ogre? If you want me, you'll have to come and pick me up off the sand like a piece of seaweed! I'm not going anywhere! I'm not running, you hear? Do you understand? No chase! The chase is off!" Geez, hope he enjoys the meal. All I can do now is get down on my knees and start saying my prayers.

Saaay, why isn't that dope getting any closer? Didn't he hear what I said? For heaven's sake, I've practically thrown myself in his mouth, and he's just standing behind that hill smacking his chops. I guess that razormouth didn't like what I had to say. Guess he didn't like the fact that his game of cat-and-mouse is off. "Hey, Dino, why don't you hurry it up already? Get it over with, would you? And lose that dimwitted expression on your face if you want to be taken seriously."

What's the matter with that Neanderthal, anyway? Something's definitely wrong with this picture. Well, if he won't come to me . . . I might as well get as close as I can and see what's going on. What do I have to lose, anyway? Just my life?

Gee, the closer I get, the funnier that guy looks. In fact, he doesn't even look real. . . . Maybe that's it—maybe he isn't real. Maybe someone erected a fake monster to scare people away. Yeah, I'll bet that's exactly what this is. That thing's eyes aren't even focused on anything. Boy, I thought it was looking at me, and it's just staring out into space somewhere. . . . I definitely have to take a closer look at this.

At World O' Wonders

I don't believe it. Do I feel foolish. No wonder it wasn't moving toward me. I was right—it isn't real. It's World O' Wonders! That's where I've heard the name Phleebhut before. My dad used to bring me here when I was kid, and that giant lizard—his name is Mog.

I was just a little tyke running around in coveralls when we used to come here. My dad and I would take the elevator to the top of Mog's head and look out at the view through his eyeball! I can't believe it—World O' Wonders! This brings back so many . . . *(sniff)* . . . memories.

But no wonder I didn't recognize the place. All the rides are gone. There used to be all sorts of fun things to do around here. Now it looks like Mog and the gift shop are all that's left. Gee . . . *(sniff)* . . . why do things have to change so much? I used to be so afraid of the double Ferris wheel. Now there isn't any Ferris wheel. Darn it, I wouldn't be afraid of it anymore, now that I'm grown up.

Gee, who are those guys? Some kind of funny-looking aliens—in fact, it looks like a whole family of them. It also looks like they've had a bang-up good time. Hmmm, maybe there's some fun stuff in that store. . . . Now who's that weasel? Oh, brother. I heard what he said. Some gem, that guy. Must be the owner, and I'll bet he's a real taker—like that Tiny jerk. We'll see about that. No telling what kind of junk he's going to try to unload on me. Too bad it won't work—I only have enough buckazoids to buy a *few* things.

Saaaay, what's in that display case—an Anterean Slime Devil? Never heard of it. I'll bet there's no such thing. He's probably got some helpless little animal dressed up like a slime devil—whatever that is—just to attract attention. Anterean Slime Devil—right. And I suppose if I open this glass case and pick it up, it'll do its slime number on me or something. Well, we'll just see . . . nah, on second thought, who

148

cares? I don't have time to mess with a fake freak show monster. I'd rather get something to eat and head out of here—and off this disgusting planet for good. I guess I'll go in and see who runs this joint.

Inside World O' Wonders

"Well, hello there, uh—Fester?—it figures. Uh, how's business, Fester? Good? Good. Yeah, I saw those suckers . . . er . . . strangers leaving with their arms full. Looks like you unloaded . . . I mean *they* loaded up pretty well. So what's there to eat in this place? . . . You *don't* sell edibles here? Boy, I can't believe it. This place really has changed over the years.

"Yeah, my dad used to bring me here years ago. It used to have the best caramel apples, hot dogs, and tater sticks of any place around. Oh, and N & Ns galore—you know, those candy-coated things that melt in your hands, not in your mouth? . . . Yeah, they came in plain or walnut. My dad and I could eat a whole bag of them, except we'd pick out all the green ones.

"So you don't carry candy or hot dogs anymore? That's too bad, because I'm really not interested in buying. . . . There's a *Monolith Burger* near here? Oh, boy! What sector? I just can't wait to sink my teeth into one of those big, fat, juicy burgers. Mmmmm, I've been craving one for as long as I can remember now.

"Well, guess I'd better get on the move—I'm starving! . . . Huh? Oh, uh, maybe another time. I really can't stay right now to look at this jun . . . uh, genuine stuff you have here. Sorry, I don't think I'm inter— Hey, where did you get that? It's a one-of-a-kind? I've never seen an Astro Chicken hat before, but I sure love to play the game. Yeah, Astro Chicken's my favorite arcade game! How much is it? . . . Oh, that much.

"Well . . . oh, Orat on a Stick! Boy, you really do have a bunch of neat-o stuff in here! You know, I leveled that dude back on Kerona . . . Kerona—you know, the planet? I'm sure you've heard of it. Well, I don't care if you've heard of it or not; I *did* do away with some guy named Orat that looked exactly like the one on the end of your stick, only he was a lot bigger. That's right; I blew him up with a can of dehydrated water. He was all brawn and no brains. How much is that, anyway? . . . Geez, your stuff is pretty expensive.

"I don't think I can afford . . . what kind of underwear? And they're all the rage? How come I've never heard of them? Who me? I'm from Xenon. . . . They're all the rage on Xenon? I sure wish I could afford all these neat things you've been showing me, but I barely have enough buckazoids for a couple of Fun Meals at the Monolith.

"Saaaay, what kind of rock is that in your display case? It's called *what*? Ohhhh, *orium*—and it's very rare? Oh . . . well I knew it was expensive, even if I didn't know the technical name. I just happen to have something here you'll be very interested in. . . . That's right, a big chunk of rare *orium!* I've been carrying this jewel around in my pocket for months now, ever since I stumbled across it in a secret underwater cavern.

"Actually—Festus is it?—I've had this stone appraised by a couple of different gemologists, and they all agree that it's very old and very valuable. Supposedly it was in that cavern for centuries. Some brave soul put it there for safe-keeping, to be given as a gift to his bride in waiting some-where—only he never returned. It's a sad story.

"Anyway, Festus, if you're interested—and for the right price, of course—I just might let you take this flawless orium sample off my hands. I do get a little worried every now and then about carrying something so valuable around in my pocket. Of course, since I'm a nomad of sorts these days, there really isn't any place I'd feel safe leaving it. You know what I mean, Festus?

"Well, Festus, what do you say? Are you interested? . . . What? You've got to be kidding, Bozo . . . uh, ahem . . . I mean, while your kind offer does indeed have merit, I'm afraid I couldn't possibly part with an orium gemstone for that price. Why, Festus, you'd be practically *stealing* it from me, ha, ha, ha. Nooo, I just couldn't accept that offer. What was that? Hmmm, I don't know. That really isn't much better than your first offer. You know, I guess this just isn't going to work. I was really hoping to buy some of your fine merchan-dise, but it looks like I'm going to leave the same way I ar-rived—with just enough buckazoids to buy a couple of Fun Meals.

Oooooo, Festus thinks he's landed a big one here! You could land yourself some big bucka-zoids—425 to be exact.

"Well, Festus, it was nice knowing you and nice doing business with you—almost. You take care, now, and keep up the good work selling all this neat stuff. You're sitting on a real gold mine, you know? . . . How much? Did you say 425 buckazoids? Hmmm, that's an offer I can hardly ignore. Let me think about this for a minute. Hmmm, 425 buckazoids. That's an offer I just can't refuse—it's a deal! For 425 bucka-zoids, the orium is yours.!

"Great, thank you very much, Festus, and here's your orium. Now, let's see. I wanted to do a little shopping before I leave. Honestly, this is the coolest store I've ever seen. I'll take that Astro Chicken hat—no, just one will do for now—and the thermoweave undies, and that awesome Orat on a Stick! Good deal! Appreciate doing business with you, Festus!

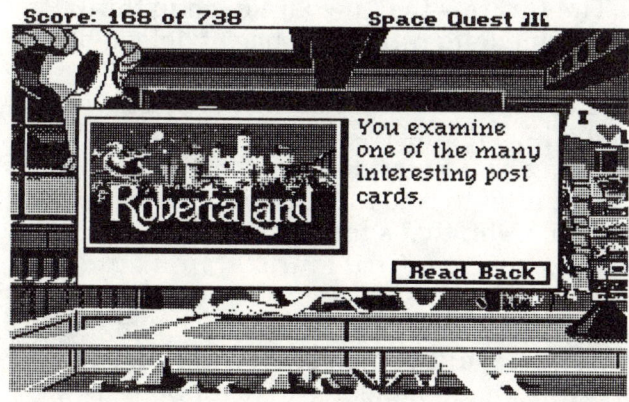

You examine one of the many interesting post cards.

Read Back

While RobertaLand may be significant to your fantasies, you'd best concen-trate on the post-card significant to your future!

151

"Say, those look like pretty neat postcards on that rack. Mind if I look at a few? Thanks, Festus, you're my kind of guy!" Hmmm, this one's interesting—but I've never heard of RobertaLand. I'll bet this place is even better than the old World O' Wonders. Gee, what a nifty, catchy name. Roberta must be a really swell gal to have such a great place named after her. I'll have to meet this lady—hope she isn't married or anything. I'll bet she wouldn't have her nose stuck in a computer all day, either—like Gladys. Oh, well. Someday I'll come face to face with you, Roberta—whoever you are—and together we'll have a blast on one of those dynamite rides at RobertaLand.

Hmmm, this is an interesting card—Ortega and its orbiting moon of Pestulon. I know I've heard of that place before. There's nothing to it but a sea of molten rock and lava—it isn't the least bit inhabitable, from what I remember. I sure am glad I subscribed to *Planetary Geographic* all those years. In fact, I still have every issue. Of course, what good does that do me now? I'll never . . . (sniff) . . . see home again anyway. Home—now there's a concept. I didn't ever have a . . . (sniff) . . . real home . . . just a couple of favorite closets to call my own . . . (sniff).

Oh, geez, Wilco, pull yourself together. This is no time to get sentimental. I have everything I need here, including a slew of buckazoids. I guess I should think about the future now, not the past. The past is gone forever. I'll never pass that way again—not ever. Never. No matter what future adventures await me. I know for a fact that I will never return to the past. That only happens in the movies and computer adventure games. The future—that's what I need to think about now. And my immediate future is a bag of delectable goodies at the first Monolith Burger I can find.

"Well, so long, Festus. It was great doing business with you. . . . No, I don't believe I'll need anything else today. You got that orium for a song, you know. Oh, yeah, I've seen the elevator. My dad and I used to ride it every time we came to World O' Wonders. Oh, right, that's too bad it's broken now. You should get that fixed—it's a great tourist trap . . . uh . . . I mean attraction.

"Oh, by the way, you didn't happen to see some big iron pumper walk through here, did you? Well, it's the darnedest thing, but evidently this metal man, or whatever he is, can make himself disappear by pushing a button on his belt. . . . No kidding. I saw it with my own eyes. In fact, I watched his footprints as he walked off, and since he was headed in this direction, I was just wondering. . . . So you haven't seen him? Well, that really doesn't surprise me. He wouldn't have left any footprints in here, you know. Better check and make sure all that orium's still in the display case, if you know what I mean. I'll be on my way now. You be sure and keep a watch out for that iron pumper, you hear? Later, pal."

Outside World O' Wonders

Whew, thought I'd never get out of that place. I really pulled one over on ol' Festus. Now that I think about it, that orium must be a dime a dozen, considering how many pieces he had in that display cabinet—hee-hee—not to mention the fact that orium must be strewn halfway across the universe on every planet. After all, I found my piece on Kerona. That was eons ago! Oh, well. Festus had it coming to him, calling those poor little visitors alien scum. What a jerk!

That little family looked like they were having the time of their li— Ahhhhh! Wh-at's . . . go-wing . . . on? . . . Haaaalp! Some-bod-y . . . help . . . me-e . . . ah-h-h-h-h-h . . . (cough, cough) . . . uhhhh . . . (gasp) . . . ahhhhgggg . . . I h-hear y-you! . . . Y-you're ch-oking me-e . . . (cough, cough). . . .

Whew! I don't believe it! I *knew* he looked familiar! That was *Arnoid the Annihilator*, and he IS after me! Vending machine fraud! I never cheated any vending machine—well, maybe just a little bit, that one in Ulence Flats—but still, is that a reason to annihilate a person? He's the guy I saw in the spaceship—the one who flashed by on the way to this planet. That jerk followed me here. He's been looking for *me* . . . (gulp) . . . the entire time!

Boy, I don't know if I can take this. He said he'd give me a few minutes to get back to my ship and then maybe he'd let me live. I don't believe him. I think he's going to terminate me anyway—that's his job. He wouldn't know how to do anything else! What am I going to do? I don't stand a chance against that . . . that . . . Annihilator! I have to think fast—I don't have a second to spare. Let's see; what are my options? I know—I could get him to follow me to the broken elevator

Uh-oh, it's Arnoid the Annihilator. What did you do this time, Roger?

ride, and then when we get to the top, I could push him over or swing a heavy pulley at him or something!

But no, that's just too risky. What if he were still alive after falling? What if he doesn't have to be completely intact to operate? After all, he's nothing but a bunch of wires programmed in a suit of armor. Hey, maybe that's it. Maybe I could reprogram him or something after making him fall. Oh, blast it—what am I thinking? I don't know *how* to program. Heck, I can barely type my name on a computer terminal, much less reprogram a murderous android.

No, there has to be another way. What about the giant snake? Yeah, I could get him to follow me to the giant snake, and . . . and the snake just might eat me, not him. Darn it, I have to think of something. Wait a minute—the pulsating pizza things. Yes! That's what I'll do! I'll lure him to that rock with those creatures clinging to it and then get him to walk under the rock. It just might work. Those throbbing things can land on his back, not mine.

At the Pulsating Pods

OK, Arnoid, I know you're around here somewhere. It's about time you ditched that disappearing act and came out into the open for all to see. Gee, he should be following me by now. I walked right by my ship to get over here. I'm sure he must have see—Uh-oh, I see footprints—I believe I have company. OK, pizzas, do your stuff. Get rid of this guy for good—that's all I ask.

"Hey, Awnoid, you hiding behind invisible shield? You some kind of cowad, you know? You talk funny, you know? You motha neva give you no English lesson when you grow up? You employ double-negative all the time in you speech? Hey, it's time you come out in the open, Awnoid. You going to twy and annihilate me? You have to come ova hea and get me foist. . . ."

It's working—at least I . . . *(gulp)* . . . hope it's working. He's coming this way. I don't think he cared for the conversation. That's it; just keep on coming. "Hey, when you going to learn to talk right, Awnoid? You think you some kind of tough guy, no?" Uh-oh, a thought just occurred to me—what if those throbbing things are harmless? What if they're nothing like the ones I remember seeing and hearing about?

What if this plan of mine is really just stupid and bird-brained and is on the very verge of coming to nothing? What if I'm going to be annihilated in about three . . . huh? What's that? What's it doing? Ahhhh, look at that thing! It's horrible—I can't look. They're even grosser than I thought! That could have been *me* being eaten alive by that horrible creature! They don't fly and land on people's backs—they're some sort of disgusting, gross, throbbing pod creatures!

Ohhh, I feel sick on my stomach just thinking about what might have been. But no, Wilco, you *can't* think that way. I've just annihilated the Annihilator—beat him at his own game. I should be proud of that. I *am* proud of that, and now I can get the heck off this nightmare of a planet and . . . saaaay, that looks like Arnoid's belt. I guess the pod creature didn't like the taste of it. Hmmmm, that could prove to be very useful to me sometime. I can think of at least a dozen times when I would *love* to have been invisible—particularly when the boss was looking for me. Even better yet, when I took an occasional nap in the closet. Yeah, this could be a revolutionary device for all janitors!

I have to get that belt—but how? I can't crawl under there—the pod will get me. What do I have here that would work? Let's see; maybe I could find a stick or . . . stick? I have a stick! I have Orat on a Stick! That's it! Orat's little mouth opens and closes—I could use it to grab that invisibility belt! I knew I made some good purchases at the World O' Wonders souvenir shop. OK, Orat, do your stuff. . . . All right, I got it! I got the invisibility belt!

Well, this has been quite the adventure since leaving that old trash barge. Now all I want to do is get in my little battle cruiser, find the nearest Monolith Burger, and chow down.

Back at the Ship

Well, I have everything I need, I hope. Too late to go back now—I don't ever want to see this planet again. Scorpazoids and giant snakes—this place has really gone downhill. Heck, I'm much more interested in RobertaLand—or rather, Roberta! Now that's something I could dream about.

OK, all systems are go. Engines are working still—thank goodness—radar is on, and we have liftoff! . . . Let's see; I need to scan the sectors here and hopefully find a . . . Monolith Burger. Ship, make tracks!

Game Points Earned

Action	Points
Buying Orat on a Stick	5
Buying the hat	5
Buying the underwear	5
Getting the invisibility belt	35
Luring Arnoid to the pulsating pods	45
Selling the gemstone for 425 buckazoids	8

Object Locations

Object	Location
Astro Chicken hat	World O' Wonders
Invisibility belt	Arnoid's body
More buckazoids	World O' Wonders
Orat on a Stick	World O' Wonders
Underwear	World O' Wonders

Monolith Burger

Mission: Stuff your gut, play Astro Chicken, and decode a secret message before leaving.
Total Points: 140

Monolith Burger

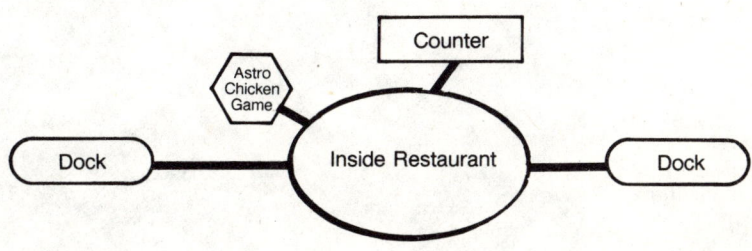

Inventory

Decoder ring

Outside the Monolith

There it is! All right, I love the sight of that platinum M. Hey, there go my fellow space questers! What difference does it make that they don't know me from Adam—they're all *my* heroes. And just think—we all like to eat at the same restaurant.

What is it with space heroes? You're always in a hurry at mealtime. Try taking it easy for a change, Roger. Make this a "fun" mealtime.

"So long, guys! I know you can't hear me, but I hope your next five years are just as good as your last five! Good luck!" It's pretty cool to see the crew of the *Enter*— Oh, it's time to dock. The Eagle has landed. Burger and fries, here I come.

Inside the Monolith

Boy, there were times when I thought I'd never see a Monolith Burger again. Oh, the feeling I get from being here—it's like a good friend welcoming me home after a long absence.

I'm so hungry I can't stand it—heck, I could eat 20 Fun Meals the way I feel right now. My stomach's about to cave in, and that wonderful burger aroma isn't helping much! Wow, look at all these great fellow travelers in here. I wonder which faraway lands they've journeyed from—and how they all happened to end up here at the Monolith Burger. Some of them are probably still on their travels—their destinies unknown, except to themselves.

They all look pretty friendly here, too. Maybe some of them could tell me all the places they've been. "Excuse me. My name is Roger Wilco, and I've come a long way alone. I couldn't help but notice that the two of you are together. Are you on your honeymoon or just passing through?

"Oh, so you do speak English? Huh? I did? Well, I didn't mean to break in line. Actually, I wasn't trying to—I was hoping to meet a friendly face. . . . Look, I'm sorry about moving in line. I'm just standing here. I'm not IN line. . . . Well, I'm not trying to crowd, either, and just who are you calling jerk, snoutface? You and Ms. Widette. . . . I'm not the jerk; *you're* the jerk . . . no, *you're* the jerk . . . no, *you're* the jerk . . . are, too . . . are, too. Blaaaah!"

Imbecile. Who does he think he is? I was just trying to be friendly. Probably thinks I was after that putrid-looking thing clinging to his arm. Ugh, don't make me gag! Oh, well. Forget those jerks. I'm here to enjoy a savory supper or two. Hey, that counter on the other side is empty—I'll just dash on over there before it gets crowded. I'm ready to chow down!

Not much to look at, but she can sure turn on the charm—with an almost hypnotic magnetism.

"Ahem, excuse me." Oh, brother. Where'd they dig up this fly with the big nostrils? Talk about something that'll make you lose your appetite. And those teeth! She looks like a walking can opener—or more like the skeletal remains of one.

"I'll have a Fun Meal and a large shake. . . . Yeah, that's fine, but I thought those were included with it. . . . Well, sure, you can add that, too. . . . Hey, what is this—rip-off city? I thought all those things were included *with* the meal. . . . Oh, well, excuse me for breathing. I don't keep up with your stupid menu every week. That's a pretty brainless way to run a business, if you ask me. Yeah, it figures. That's the standard answer these days—you don't know; you just

work here. Just give me my food, OK? Here are your buckazoids. . . . Next time I'll try not to overload your brain by asking so many mind-boggling questions. . . . You're welcome."

Gee, I don't know about this—my impression of the
Monolith is faltering quite a bit after this last episode. Maybe
they've changed management or something. Oh, well. What
do I care? I have these yummy vittles to polish off—and
that's my first priority. So what am I waiting for? "Over the
lips and past the tongue. Look out, stomach; here it comes
. . . (chomp, chomp, chomp) . . . (slurp, slurp) . . . (chomp,
chomp, chomp) . . . (sluuuuurrrrp) . . . (chomp, chomp,
chomp) . . . ouch!" Hey, what the heck kind of burger is this?
I thought they were supposed to get rid of the bones
before. . . . Wait a minute. This isn't a bone—it's a decoder
ring! I forgot about the Fun Meal prize. Wow, I've been wanting one of these for a long time. After all, how can I possibly
compete with the likes of Indiana Jones and company if I
don't have a top-secret decoder ring?

Gee, first I buy all that great stuff in World O' Wonders,
and now I have a decoder ring. Who cares if it came from a
Monolith Fun Meal—it's still the real thing. Boy . . . (chomp,
chomp) . . . yummy . . . (chomp, chomp) . . . meal . . . (slurp,
slurp, slurrrrrp). . . . I could eat another whole Fun Meal. I
have enough buckazoids to buy a hundred meals if I want
them. As a matter of fact, I think I will get another bite to
eat. . . . Oh, hey! It's Astro Chicken! If this isn't my lucky day,
I can't wait until the real thing comes along! I can't believe
it—my most favorite arcade game in the universe. Forget the
Fun Meal, Jack—I'm ready for some arcade action!

At the Astro Chicken Game

Boy, Astro Chicken is my favorite cartoon hero of all time. I
used to spend hours playing this in Orbital Station 4's game
room. I was the best Astro Chicken player around—no one
could beat me. I even landed over a hundred chickens in a
row one time. Maybe it's been a while since I've played—
considering what I've been through lately—but I'll bet I
haven't lost my touch.

Don't give up on your fine-feathered friend too soon. Persistence has its rewards.

I really used to rack up the buckazoids on Orbital Station 4. Darn, what I would give to run bets with those guys again. Gee, what memories. I was darn good—a real gamester. Well, let's see how well old Roger Wilco fares this time around. Get ready, Astro Chicken—you're about to hit that landing pad a dozen perfect times!

Let's see; the buckazoid goes in this slot—and we're off! OK, slow and easy now . . . a little to the left . . . oh, get back over there! . . . Hey, wait a minute; this thing's going too fast. Where's the speed control? Oh, there it is. Why didn't I see that to start with? OK, I'll just insert another buckazoid—and here we go!

Great, it's a lot slower this time. Now I can do my stuff and land this turkey—oh, ha-ha, excuse me—land this *chicken* on the pad. . . . What? What the. . . . Hey, what's wrong with this game? Something isn't right. No way I could have lost my touch in just a few short months. Well, I'll try it again. That's two buckazoids I've lost so far, but I'm not giving up until I land at least 50 chickens. OK, here we go again. I'll insert another buckazoid, and . . . whoa! Move over there! MOVE! . . . Get over there, darn it! On the *pad*, you stupid duck, on the *pad*! . . . Shoot! I don't believe this.

Something's wrong with this game. I haven't landed a single Astro Chicken. Maybe I should've put on my new hat for good luck. Oh, well. I'll try it again—here goes my fourth buckazoid. . . . Ooooo, GET OVER THERE. . . . Go, go, GO! . . . All right! I landed an Astro Chicken! Boy, that gets my fever burning, even if it is only one point. OK, here we go again. . . . Move over . . . over . . . OVER, darn it! Move to the left, you dope! Move to the left! . . . One more buckazoid. . . .

Blast it! I've spent nine buckazoids and I've only landed one Astro Chicken. Stupid cheating game—this thing is rigged! Darn it, I've never been so mad in my life! OK, one more buckazoid and that's it. I'm not throwing away any more of my money on this rigged-up casino gimmick. This is highway robbery! Well, here goes my tenth and final buckazoid—probably down the drain again.

Just one more time—my tenth and final buckazoid. OK, get over there. . . . GET OVER THERE! . . . Get on the *landing pad*, birdbrain! . . . *Darn* it! Well, that's it—I'm not playing again. I've had it up to here with this machine. Someone ought to reimburse my entire ten buckazoids! . . . Saaaaay, what in the world is that? I knew something was wrong with this game—just look at this messed-up screen. . . . Hmmm, it almost looks like a . . . a . . . *secret code!* I'd better get out my top-secret decoder ring and find out. This message could be of universal importance!

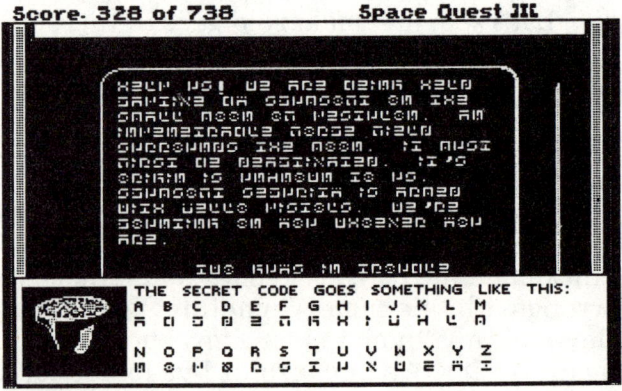

Yrros, ll'uoy evah ot erugif siht eno tuo no ruoy nwo! Llew, yako—fi s'ti yllaer taht tluciffid rof uoy, ll'ew evig uoy a yneet-yneew tnih. Kool ta eht xidneppa.

OK, ring, do your job and decode this message. . . . It is a secret message. My ring really did the decoding trick—well, sort of. It looks like I have to match up letters of the alphabet to those funny-looking shapes. That could take forever, but who cares? I just have to find out the meaning of this message!

Gee, I bet this has to do with some top-secret interstellar operation. I'll bet *Two Guys* is just code for some high-level Space Fleet agents. Why, this could even involve—espionage. I'll bet the Two Guys put their message in Astro Chicken because no one would think of looking there in a million years. Most of those guys—the major spooks and all—probably don't even know what an Astro Chicken is. Gee, what if all these people in here are really secret agents just pretending to be customers? What if this actually isn't a Monolith Burger restaurant? What if it's just a very clever front?

I must be pretty smart, since it didn't fool me. Well, maybe it did just a little at first. . . . A thought just occurred to me. What if the Two Guys have heard of me before—already knew of *my* inimitable reputation and the way I single-handedly infiltrated both the Sarien ship and Sludge Vohaul's asteroid? What if I was meant to see that message—just me? That's it. They knew that I love Monolith burgers and that I'm an expert at playing Astro Chicken, so they set the whole thing up for me to find their message. They knew I would eventually come to Monolith Burger and that I wouldn't be able to resist playing a few rounds of Astro Chicken.

Those guys are brilliant—their plans have gone off without a hitch so far. Good thing they stopped at ten games, though. Any more, and I would have walked off without ever playing Astro Chicken again. They *knew* I'd become frustrated and want to keep playing if I started losing over and over again. What a stroke of genius.

Well, since the Two Guys have given me the responsibility of rescuing them, I can't let them down. What a job. I just hope I can handle it. I guess I should get ready for action. . . . Now to find Pestulon—but first, the force field that surrounds it must be destroyed. Hmmm, one of the postcards I read at World O' Wonders featured Ortega—and its orbiting moon of *Pestulon.* Ortega must be the source of that force field. Wellll, it looks like Roger Wilco is taking a trip to Ortega, the planet of volcanoes!

Oh, gee. I forgot which door I came in. I'll try this one next to the Astro Chicken game. . . . "Hey, watch it! How the heck was I supposed to know the dock was being used? Listen, jerk—I don't think you know who you're talking to. I'm Roger Wilco, and I've just been given a special assignment to . . . uh . . . to clean out all the docks in this restaurant! So there! I had a right to be in there, you dunce!"

Whew, that was close. I almost gave away my secret mission. I'd better be careful about what I say from now on. As far as everyone else is concerned, I'm still Roger Wilco the janitor. OK, it looks like my ship is docked on the other side. I'll just make my way over there right now and get this show on the road . . . er . . . I mean outer space!

Back in the Ship

OK, Wilco, there are two very important lives depending on you. Wow, I can't believe this. I can't imagine that they could possibly want me, a janitor, to be a part of . . . of . . . the Space Fleet Sub-Rosa Service. Yeah, that's it.

I'm ready to go—engines are fired, and radar is on. . . . Takeoff is commencing smoothly—dang, I did a good job on this ship for a nonengineer—and we have liftoff! . . . Now to set the course for Ortega . . . OK, the course is set for that red-hot planet . . . and light speed is engaged! I wonder what Captain Picard would think.

Game Points Earned

Action	Points
Buying and eating a Fun Meal	10
Decoding the secret message	20
Finding the decoder ring	10
Finding the secret message	50
Successfully landing ten Astro Chickens	50

Object Locations

Object	Location
Decoder ring	Fun Meal

The Planet Ortega

Mission: Find a way to destroy the force field generator and escape the planet safely.
Total Points: 80

The Planet Ortega

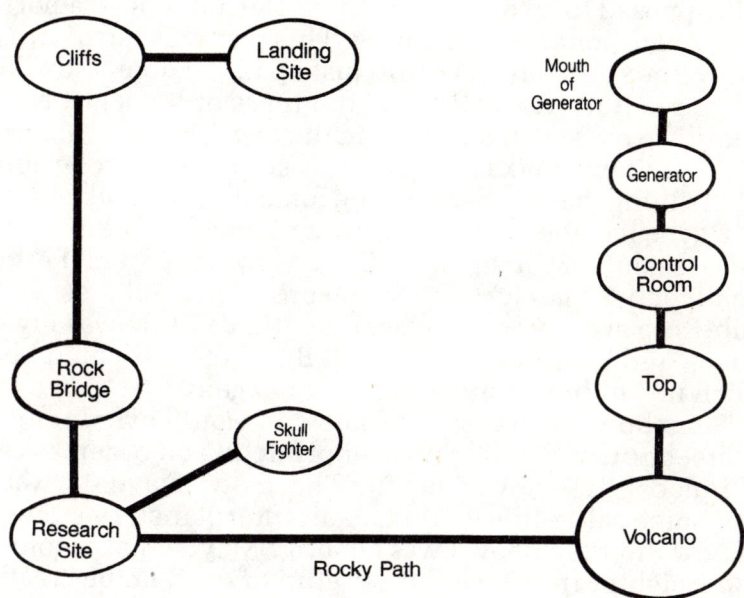

Inventory

Detonator
Pole

Landing Site

So here we are at Ortega, the red-hot planet. Well, that's OK.
We Sub-Rosa Service agents have to brave all sorts of ele-
ments and natural calamities—it's part of the job. If we
couldn't take it, they wouldn't recruit us to start with—and
I'm convinced I've been recruited.

Boy, red-hot is an understatement for this planet. How
am I supposed to go out in that heat? Darn it! I need a portable air conditioner or something. This planet is too darn
close to the sun. Maybe I could just open the door and take a
peek to see if it's really that hot out there or if it's just me.
Hmmm . . . oh, gee. It's hot all right. Now what am I going to
do? If I can't even take a step off this ship, I'll never be able to
find the thing that's generating the force field around Pestulon.

Hey, what about that underwear I bought? Didn't Fester
say it was for very hot places? Saaaay, I wonder if ol' Fester is
actually more than just a gift shop proprietor. I'll just bet he's
a Sub-Rosa agent under cover. Yeah, that's it! He was my
contact, and I didn't even know it. But then how could I have
known it? I didn't even know *I* was an agent!

Somehow *they* must have known I would eventually go
to Phleebhut, and Fester was simply acting on orders when
he "innocently" showed me the Thermoweave underwear.
Wow, some setup this is. Talk about interplanetary strategy.
But how did they know I was coming? Maybe that whole
trash freighter fiasco was just a setup to see if I'd be creative
enough to escape it. Yeah, I'll bet that's it. Come to think of
it, maybe everything that's happened to me has happened
because of the Sub-Rosa Service. Boy, they sure know how
to test a guy. I could really be mad about some of the things
I've been put through—if my theory's correct. Now my assignment, obviously, is to find a way to destroy Pestulon's
force field and then rescue those Two Guys. Soooo, the plot
thickens.

Oops, before I go, I'd better remember to wear this spe-
cially made Thermoweave underwear. Gee, who knows?
Maybe it's laced with cyanide just in case I'm captured by
the enemy! But then what? I wouldn't know what to tell
them anyway. Oh, I've got it—I'd give them my name, rank,
and serial number, and then refuse to cooperate any further.
. . . Gosh, but I don't have a serial number, and as for rank.
. . . Oh, well. Guess I'd better play it smart and not get

caught. That's the best way to avoid having to answer any taxing questions.

Gee, it *is* hot out here. It's so hot I'll bet the corn on this planet pops before it's been husked. I'll bet the fish fry up before they've even been caught. And I'll bet they feed their cows Hershey bars and then milk them for hot chocolate! Ah, ha-ha-ha! Hee-hee, haw-haw, hoo-hoo! Darn, if I'm not funny. . . . And the next thing I know . . . ha-ha . . . I'll be seeing little pitchforks and pointed tails on little d-d-d-DEVILS.

Gee, it's not so funny anymore. What if there really *are* little devils running around here? I mean, this place is like a furnace, and anything that hot breeds d-d . . . bad things. After all, Hades by any other name would be as hot! Ohhh, I don't even want to think about it. OK, Wilco, don't go and get yourself all worked up now. I haven't seen the first sign of any little devils yet—not even the Tasmanian kind. And I haven't heard the wails of suffering souls yet, either.

Oh, gosh. Why in the world did they pick me for this assignment? I've had enough assignments. And why hasn't anyone contacted me about any of this? I g-guess that's the way the Sub-Rosa Service works. I g-guess that's the way it is when you're important enough to work for The Company. Well, n-now that I'm here, I'd best make my way around as quickly as possible and get off this . . . this volcano!

A little fancy footwork might help for the moment, but don't expect to just walk across this path when you return. If you're smart, you'll jump on any equipment you happen to find lying around.

Saaay, what's that over there? Hmmm . . . y-y-yikes! It's a big skull! Ahhhh! Ortega really is he— Oh, ahem. It's just a spaceship. JUST A SPACESHIP? Wait a minute. You mean to tell me I'm not the only one on this planet? Oh, g-g-great. All I need is to be stuck here with a bunch of people who paint their spaceships like skulls. Gee whiz, how do I get myself into these messes? All I know is those Two Guys must be awfully important for The Company to put me through this. I sensed that from the message they planted in the Astro Chicken game. I suppose it's worth it for me to risk my life to save them. No telling what kind of rewards I'll get, too.

I don't think I'll hang around that skull ship—its owners are around here somewhere—close by, no doubt. No telling what they're up to. Gee whiz, it is an ugly spaceship, and it looks like it's poised for battle. Dollars to doughnuts it's some sort of fighter ship, and I'll bet the jerks that fly it are from Pestulon. They're probably the same ones holding the Two Guys, in which case, they mustn't see me! They'd probably shoot me with one of their jelly . . . er . . . jello pistols, and then I'd *never* be able to rescue the Guys. I'd better go the other way—to the west. I don't want those creeps finding out I'm here. At least my ship is out of their view—at least I *hope* it is.

Crumbling Bridge

Hmmm, I see a narrow sort of pathway over there. It looks like it's about to crumble to bits, though. Surely there are other paths around here, too. Oh, well. I have to cross it. I don't have any other choice right now.

Gee, it i-is a little sh-akey! Ooooo . . . whooooooa . . . yeeeeeowww! Boy, that was close. That darn thing could give way any moment. This planet's so shaky, and with all the molten rock and lava—gee, it's one big earthquake! One little slip, one little wrong move, and I'll be swimming in a sea of red-hot lava! I'd better hurry—but hurry with what? I don't even know what to do yet!

Saaay, what's that noise coming from behind those rocks? I thought I heard voices. And what's that little click-click-click, like a pinwheel blowing in the wind? I'd better check this out. Maybe I've discovered the crew of the skull fighter.

Research Site

I knew it! It's them, and they're definitely armed with some sort of pistol—probably those jelly—I mean JELLO—pistols the Two Guys mentioned in their message. Gee, I wonder what they're doing? All that equipment and stuff—what in the world is it? Must be some kind of research project.

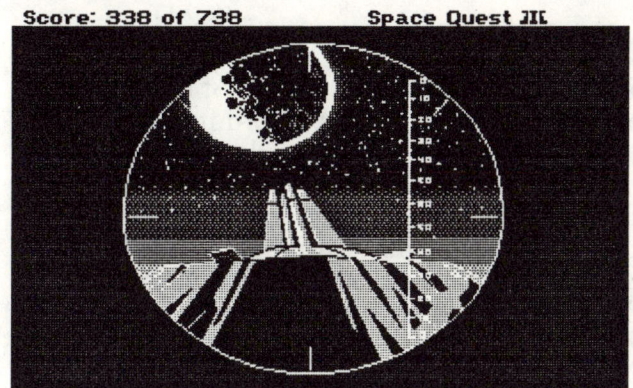

Score: 338 of 738   Space Quest III

Those pestilent Pestulons obviously think they're safe now that the Force is with them. Maybe you can find a way to burst their little bubble.

Hey, they're leaving. I'll just give them a little more time to get back to that ship of theirs. . . . Well, it looks like the coast is clear. There goes the numskull fighter, off into the clear red yonder. Hmmm, now what's all this stuff? A telescope! I've always wanted one of these. It looks like a super-duper powered one, too. Wow, I'll bet looking at Pestulon is almost like being there! Let's see; I'll take a quick look before moving on to the more important task of finding. . . . Hey, it *does* give a great view of the moon—and that's not all.

What in the world is *that* thing? It looks like some sort of . . . *generator!* That's it! That's what I'm supposed to find and destroy! Now, if I can figure out exactly how to get to it—and how to get rid of it. It's huge! I'd need a thousand sticks of dynamite just to make a dent in that thing. There must be something around here I can use.

I wonder what's in that crate over there. Let's see; can I pry this open with my bare hands? Got it! Oh, what's this— some sort of—what's it called—detonator? I'll certainly hold on to this little jewel. I wonder how powerful it is—hopefully powerful enough to blow up that generator. And hopefully, I won't get myself blown up in the process.

What's this funny-looking thing? Oh, I know; it's an amenom . . . anenom . . . amenon. . . . It's a thing that tells how fast the wind is blowing. I never could say that darn word. Who comes up with names like that, anyway? Why not call it a Wind-Speed Wheel, or Pin-N-Pole Wind Reader, or something simple like that? Judging by the angle of that telescope, it looks like I need to head due east.

Bottom of the Generator Mountain

Gee, there's nothing around here but a bunch of big rocks and a lot of fissures in the ground—and they don't look like they're going to last forever. How in the world was *anything* ever built on this horrible planet? Maybe it was once very different here. . . . Nah, I doubt it. This place looks like it's been self-destructing from day one. Hmmm, now that I'm as far east as I can possibly go, there must be some path leading up to the top of the mountain—I just can't see it.

It's really difficult to tell which way to go, since all these rocks are so similar. The ground is really shaking here, too. I really would like to find this pathway I'm looking for—or whatever it is I'm looking for—so I can get on with this task. Oh, wait a minute—I may have found it. . . . This is it—finally. Now I can get to the task at hand, and . . . hey, a stairway. Wow, would you look at this place? It's huge!

Generator Room

That looks like a ladder in the distance. I'm sure it leads to the mouth of the generator. Now, if I can only find my way through this maze of a room. What in the world is all this equipment, anyway? It must have cost a fortune to build—and operate. Surely there used to be a lot of people here operating everything. I guess they all had to leave when the planet started falling apart.

I'll bet the force field is operated from Pestulon now. Gee, I wonder what backup plans they've made when Ortega finally gives out. For that matter, it seems like the moon of Pestulon would no longer have anything to hold on to. I wonder—will it be hurled into space? Oh, well. They're going to get a chance to use those backup plans a lot sooner than anyone thinks, because I'm going to blow this baby up.

At the Generator

Finally, I made it through that maze of equipment. I guess it wasn't so bad. At least I had a clear overhead view of where I was going before I went through the maze. OK, now to climb the ladder. . . . Wow, look at this generator—it's huge! It kind of gives me the creeps being here, knowing I could blow up with this absolutely tremendous piece of machinery. It looks like a pretty steep climb up to the top of it, too. Boy, I knew being a member of The Company had its dangers, but this is carrying the idea of occupational hazards a bit too far!

Might as well get to it now. I'll have to crawl very slowly—one leg at a time. Gee, this generator thing is kind of slippery. I'd hate to . . . whoa, whoa . . . lose my balance and fall! You're almost there. . . . Whew, made it! That's a pretty steep hike. I can imagine what it's going to be like climbing back down—and in a hurry, too!

OK, what's inside of this thing, anyway? Uh-oh, I'd better not lean over too far, or I'll get a closeup—and permanent—view of its insides! The only thing I can do now is throw the detonator down its throat and hope it doesn't blow up right away. Here goes! . . . I don't hear anything. . . . That's a good sign for me, anyway—I think—at least as far as my life is concerned. I just hope this doesn't mean that the detonator isn't working or that it isn't strong enough. Oh, gee. That would mean I'd be back at square one!

I'd better climb down from here now. By the looks of this place, generator or not, there won't be any solid ground left . . . woooo, almost slipped again. OK—thank goodness—I made it back down to the base of the generator. Now to climb down the stairs, make my way back through the control room, and go down that footpath—fast.

Uh-oh, what's that rumbling? The detonator—it must have worked! Great, mission accomplished. At least I'm away from the generator now. I can't believe it didn't blow up the entire planet, but, of course, that makes sense. I didn't stop to think of the consequences for me if Pestulon lost its orbit. I'd lose Pestulon!

Right now I'm about to l-lose my f-f-f-FOOTING! Yikes, get me out of here! I've got to get back to the ship, darn it, and the only way back is across that crumbling bridge.

Back at the Bridge

I'll bet it isn't even there anymore. . . . Here's the research site, so the bridge is just past those rocks over there. . . . Oh, great. I knew it. The bridge is gone!

What am I going to do now? This place is falling apart fast! I think the detonator had more of an effect than I originally thought, but that won't do me any good now. If I can't get off Ortega, there's no point in destroying the force field. No wonder those comrades parked their ship on *this* side of the bridge—they knew what would eventually happen.

Now I'm stuck here—doomed to a very short life on this godforsaken planet. I'll bet my Thermoweave underwear won't even hold out for long. I'm doomed. Doomed. DOOMED. Why did this have to happen to *me*? What did I ever do to deserve this? . . . (*sniff, sniff*) . . . ohhhhh, wah haaah haaah . . . (*sniff, sniff*) . . . I'll bet they wish they'd never even recruited me as a Company man . . . (*sniff, sniff*) . . . or even worse, they just wanted someone disposable to shut off the force field so *they* could go in and do the rescue!

I'll never be a good Sub-Rosa agent . . . (*sniff, sniff*) . . . but wait . . . (*sniff*) . . . a minute. I'm not being a good agent *now*. I'm not. A really good agent wouldn't sit down and start crying like this. A really good agent would find a way to cross that gaping hole in the ground. A really good agent would know that even if he didn't succeed, at least he made the best darn effort he could.

Gee, I sure hope no one saw me just now—on a secret camera or something—that would be just my luck. I'd be so embarrassed! No, I've got to find a way to get over that hole. If only I could jump, but it's just too wide. Maybe I could find something to help me jump. Wait a minute. A pole! I used to pole-vault back in high school. All right, I'd forgotten about that! We used to pole-vault all over the place—over cars, bushes, swimming pools! We even pole-vaulted over the principal one time! And I know exactly where I can find a pole on this planet—at the research site. There was one attached to that . . . wind thing. Gosh, I sure hope it's long enough and strong enough. It's my only hope.

OK, I've got the pole, but I don't have much time—the ground is really beginning to cave in under my feet. I haven't done this since the eleventh grade, so I hope I haven't lost my touch. Well, I'll just take a deep breath, get a running start, and here . . . I . . . gooooooooo! Ahhhhhhhhh! I made it! What a relief! Pestulon, here I come!

Back at the Ship

OK, now calm down, Wilco. I'm safe for the moment, and I'm almost off this fireball. Engines are go, radar is on, and we have liftoff. . . . Whew! I've cleared the planet, thank heavens. For a while there, I didn't think I'd live to see another day. It just goes to show that luck can change with a little common sense and courage. Now to plot the course for Pestulon. Hmmm, I don't see it yet. What's the problem here—didn't that detonator work? I could have sworn . . . oh, there it is. Well, it's about time. That little moon is still pretty well hidden. It sure took my radar long enough to find it!

OK, just stay put, you Two Guys. Roger Wilco is on his way.

Game Points Earned

Action	Points
Getting the detonator	10
Getting the pole	10
Looking through the telescope	10
Pole-vaulting across the crevice	20
Putting on the underwear	10
Throwing the detonator into the generator	20

Object Locations

Object	Location
Detonator	In a crate at the research site
Pole	On the anemometer at the research site

The Moon Pestulon

Mission: Make your way into the ScumSoft building unnoticed, rescue the Two Guys, and then beat Elmo at his own game. Survive the "star wars" that ensue once you escape Pestulon.

Total Points: 270

The Moon Pestulon

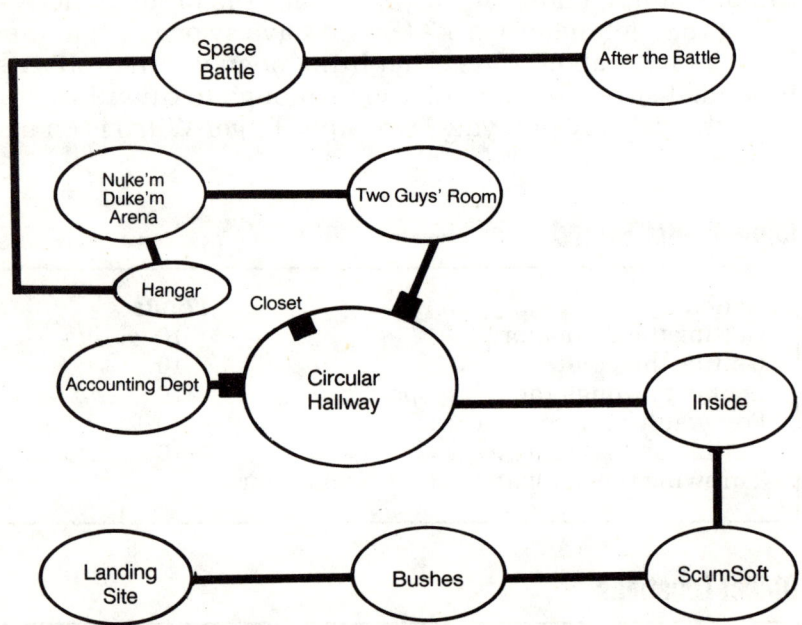

The Accounting Department

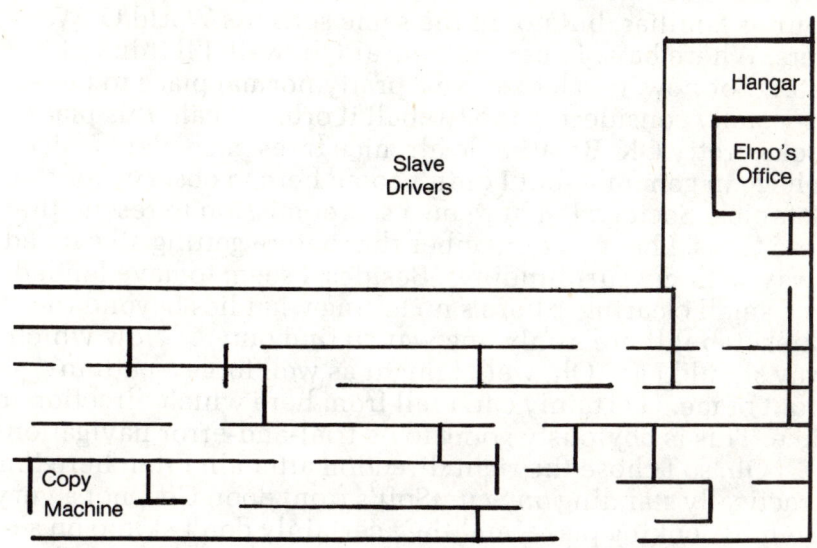

Hangar

Elmo's Office

Slave Drivers

Copy Machine

Inventory

coveralls
keycard
Mr. Garbage
photocopy of picture
picture of Elmo

Landing Site/ScumSoft Entrance

So, this is the moon of Pestulon. That name, *Pestulon*—it sounds familiar, but not in the same sense as World O' Wonders. Where have I heard it before? Oh, well. I'll think of it later. For now, this looks like a pretty normal place to me— especially considering the fireball it orbits. Yeah, this place looks pretty OK. Breathable air, nice trees, nice plants, nice foliage in general—but I didn't come here to observe for the Botanical Society. I'm here on a secret mission to rescue the Two Guys. I have to remember that before getting all carried away with my surroundings. Besides, I seem to have landed in a small clearing. There's no telling what lies beyond the thicket, but there's only one way to find out. . . . Now which way should I go? Oh, well. I might as well forge a path in front of me. I certainly can't tell from here which direction to take. This is obviously going to be trial-and-error navigation.

Oh, so I chose the right direction after all. From here I'm practically standing on ScumSoft's front door. Gee, not a very friendly looking place, and they certainly don't skimp on security. This looks more like a national armory than a software manufacturer. Hmmm, now the question is how to get by those guards unnoticed.

Score: 408 of 738　　　　　　**Space Quest III**

This would be a good time to do a disappearing act— for real.

Maybe I could sneak up on one, punch his lights out in one swift blow to the neck, and then drag him off into the woods and steal his uniform. That should work—I've seen it a hundred times on TV. Gosh, though, I could be here all day waiting for one of those guards to come this way, and then what if the rest of them saw me do it? Yikes, I don't like the sound of that—it's too risky.

I can't blow this mission now, not when I've come this far after battling such horrendous, death-defying obstacles. No, I have to think of another way to get in there. If only I were invisible. INVISIBLE? What am I talking about? I *can* be invisible—with the invisibility belt I stole from Arnoid the Terminator! Heck, if *he* can do it, *I* can do it.

Boy, it's a good thing I carry all this junk . . . er, *equipment* . . . I've picked up along the way. I just happen to have that handy-dandy little contraption right here on me. Let's see; just snap the belt into place, and then I guess I just press the On button. OK, here goes.

Gee, I don't feel any different. I didn't hear any strange noises or anything, and when I hold up my. . . . Ahhhh! My hand! It's gone! . . . I'm gone! Oh, my gosh. It's working. This is kind of unsettling, though. I'm glad I don't have a mirror. If I did and I looked at my reflection in it, I wouldn't see any-thing—which means I would look like a g-g-g-ghost! Or even worse, a vampire! They don't have reflections, either!

Oh, brother, Wilco, you're really losing it. I've got to get a grip on myself and get on with this operation. Gee, as tempted as I am to play tricks on those guys—kick dirt in their faces and grab their weapons—I'd better just get the heck inside that building and find the Two Guys. They could be in danger. I'll save the sand kicking for when I leave. Well, here goes.

Hey, I'm walking right by the guards with their jello pis-tols, and they don't even know it! I'm not even making foot-prints on the ground! This is great! I'm in! I'm safely inside the ScumSoft building, and those bozos don't even know they've been had. Eat dirt, jerks.

Wow, I see so many implications for this belt—*my* belt, since I found it. Just think of all the fun I could have on Xe-non. I could play so many tricks. . . . Hey, what the . . . wh-what's happening? Oh, no! The b-belt—it stopped working! Now what the heck is wrong with it? I have to have this belt! I can't just walk around this building without a disguise! Oh, gee. I don't believe it—the battery's dead. Well, that's just peachy. Maybe I could use something to recharge it. I'll see what it says on the battery compartment—but I have to hur-ry! Someone could spot me here at any moment!

It says, "Uses four Ronko AAA nonrechargeable batteries only. For replacement, mail 6 buckazoids, plus 2 buckazoids for postage and handling, to the Ronko Corporation." Oh, no. I should have known. No one else would have come up with this kind of gadget—it's vintage Ronko. And, of course, it only runs on their own cheap batteries. I'll never learn. Oh, well. At least it got me inside this building, but now what? One thing's for sure; I can't go out the way I came in.

OK, be cool, Wilco. This looks like an elevator door. I don't have a choice now—I'm going to have to enter the elevator and figure out where to go from there. I sure can't go back up the stairs—they lead to where the guards are, and this time, they would all see me. I'll just hope and pray no one is standing guard behind the elevator door. Here goes.

In the Circular Hallway

No one's here, thank goodness, but what the heck kind of elevator is this? This is the strangest-looking thing I've ever seen. Whoever heard of a circular elevator? Circular walkways, sure, but an elevator? This doesn't make any sense. Where are the buttons? How come there are no floors listed? Do you mean to tell me I have to *walk* to get where I'm going? This is the dumbest thing I've ever seen. What a bunch of lazy good-for-nothings these people must be.

So I have to walk. . . . Round and round and round she goes. Where she stops, nobody knows. . . . Geez, a fellow could get dizzy in this place. This *elevator* just goes on forever. . . . Say, was that actually a door I saw back there on the left? Maybe it's the way off this crazy thing! I'll just turn around and go back the other way. . . . Oh, I feel like I'm going to throw up. If I don't find that door . . . whooooo! There it issss. Ohhh, I've got to sssstop. Maybe if I slow dooooown. . . . Oh, my poor stomach. There's the door again.

OK, go slow, Wilco. Ouch! I bumped my head. Shoot, it's hard to get centered in this hallway . . . elevator . . . whatever the heck it is. Try again. . . . Ouch! I can't even get through this door. . . . Got it! Boy, that's one heck of an elevator ride—I'm still spinning. It's worse than the Tilt-a-World I used to ride at World O' Wonders.

Gee, speaking of World O' Wonders, now that I halfway have my balance back, I'd better remember my objective. I'm here for a *reason.* Saaay, would you look at this? Well, I'll be darned! No wonder I spotted that door right away; I can smell a janitor's closet a mile away. And look at what I found lying on the floor! A genuine, bona fide, for-real, authentic pair of coveralls! I don't believe it. I couldn't have planned this better myself. In fact, I couldn't have planned it *at all* myself. I'll bet a Company man just happened to leave these coveralls here by "accident." This proves it. If there's any disguise that's perfect for me, it's a janitor wearing coveralls—and they knew it!

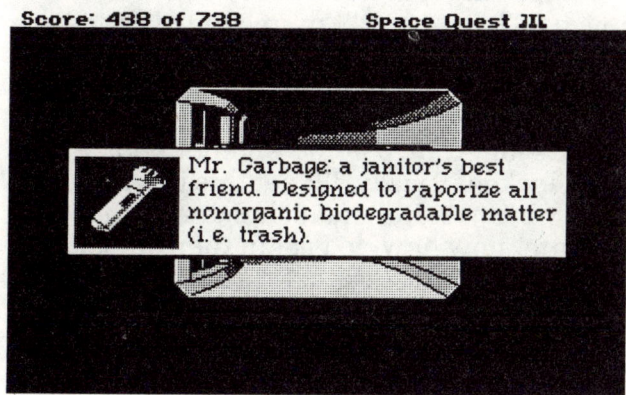

It's Mr. Garbage, the trash vaporizer! He's never seen a bucket of trash he didn't like.

I wondered how they expected me to sneak in to Scum-Soft. Well, this says it all. They had no way of knowing that I'd run into some crazy *toiminata* on Phleebhut and that I'd end up getting away with his invisibility belt. That's why they left these coveralls here for me to find. It's the perfect cover! This is great! Boy, do I feel good. Hey, what's this thing in the pocket? Not another Ronko gadget, please. Geez, I can't seem to get away from these things no matter where I go. Hmmm, this is a gadget all right, but what does it do?

Oh, gee. It's a Mr. Garbage—a trash vaporizer. I've heard of these before. In fact, they're not supposed to be half bad. Maybe it's good at vaporizing jello—as in when I meet up with the Two Guys. After all, they're supposed to be trapped in jello—at least according to that message they left in the Astro Chicken game. I'm sure that's the purpose of this little vaporizer—it's all part of the plan. I'll just put it right back in the pocket where I found it and then put myself into these coveralls. Aha, I knew it—they're a perfect fit! This really seals it. I'm in! I'm a Company man!

I can't wait to tell everybody back home. Doggone it. I can't tell a soul. I can't blow my cover, not even after this mission is over. I'll probably still have to use the janitor cover—it certainly fits, and no one would ever be suspicious.

Gee, what time is it? I've been standing around here forever. I have to go back to the circular hallway and find the Two Guys. They're counting on me! I'm their only hope. OK, Two Guys, just hold tight. Roger Wilco is on his way.

Hey, I'm not even getting sick now that I'm back in this circular hallway again. This is great! I'm so happy I can feel the lightness in my step. In fact, I'm feeling so good right now I could easily add a little skip to my strut. In fact, I could skip *and* whistle a happy tune all at the same time right now! Maybe I'll just do that—oh, wait. There's another door!

Shoot, this one's locked. Saaaay, I'll bet this is where they're holding the Two Guys. Dollars to doughnuts that's where they are. Hmmm, now how do I get in there? Looks like I'm going to need a keycard—again, for the thousandth time. I've never seen something so . . . so . . . *universal* in all my born days!

Well, I didn't see any other doors before this one. Maybe I'll find another one if I keep looking. Wow, this is getting exciting. I can't wait to try out my janitor disguise on some of the locals—if I ever see any, that is. Where the heck is everyone? Anyway, I'll be heehawing to myself so loud when they fall for this getup they'll probably be able to hear me laughing in my head.

Hey, there's another door just ahead. Ouch, darn it! These doors are the hardest things to enter. OUCH! If I bump my head into that wall one more time! . . . Boy, I'm going to have a few choice words to say to the ScumSoft people about this ridiculous hallway design. What kind of place is this, anyway? It's even worse than Vohaul's asteroid!

In the Accounting Department

OK, Wilco, try again. All right, the door's opening. Looks like I get to do the disguise part of my mission after all. Hmmm, I wonder if I should speak to anyone. All of them seem to be minding their own business. Speaking of business, what is this place, anyway—a white-collar clearing house? I've nev-

er seen so many starched white shirts with red ties in my
life—and they all wear the same kind of glasses! I know what
this place is—it's an accounting factory. I'll bet they mass
produce from one of Vohaul's trashed insurance salesman
clones—ha-ha!

Oops! What's wrong? Uh-oh, I must have aroused suspicion somehow . . . *(gulp)* . . . they've spotted me as an intruder! Y-y-y-yikes, wh-what's that thing on the ceiling? It's
coming after me! I'd better hightail it out of here right now,
on the double! Whoooooooaaaaa!

Whew, that was close! But how did they know I wasn't a
real janitor? What could I possibly have done to make them
think I was an impostor? And gee, why aren't they coming
after me now that I'm back in the hallway? Hmmm, these are
all very important questions—good questions, if I say so myself. Unfortunately, I don't have a clue as to their answers!
Maybe I should get out my vaporizer and just zap them all,
one at a time. Nah, there are too many of those guys in there.
I'd never get away with it.

Besides, I don't even know if this vaporizer works on
people. I don't see any kind of instructions for zapping people—not that they would actually include them. Oh, wait. I
didn't notice the fine print when I was looking at it before.
Hmmm, "The Ronko Mr. Garbage. Perfect for vaporizing
light trash. Works particularly well on paper products and
food. Hold canister 10 to 12 inches from debris to be vapor-
ized. Aim directly at debris and zap in one swift motion. Not
a play toy. Keep away from children."

Hmmmp, not a play toy my foot. I plan to have a *lot* of
fun before it's all over with. Saaay, a light bulb just went off
in my head. Yes, of course! Why didn't I think of this before?
My Space Fleet associates did; that's obvious. No wonder
those accounting nerds were so suspicious of me. I'll bet I'm
supposed to *zap trash*, mind my own business, and keep my
mouth shut. Obviously, janitors get no more respect on Pestulon than in any other place. OK, time to try again. I sure
hope I'm right about this trash-vaporizing deal. I wouldn't
want to be dubbed an impostor the second time around.

Well, everything's calm so far. Those tubs of lard are still
over there drinking water. Looks like they need to drink
more water and a lot less beer, judging by the size of their
guts. I'd better zap that first trash can before anyone gets sus-

picious. Maybe I'll just pretend I'm zapping the accountant
rather than the trash. Yeah, that's it! Like I said, not a play
toy my foot. Hey, Pencilhead, amortize *this*. Bzzzzzz.
Whoops, made a mistake! Too bad, pal—you're outta here!

Boy, I can see this is going to be fun. All I have to do is
pretend I'm zapping accountants while I'm zapping trash.
That way I won't miss any trash cans—and I'm sure the consequences of missing a microscopic fuzzball around here
would be deadly. There's another can—heck, there are cans
all over this room! Hey, Pencilhead Number 2, I hope you realize your realization report really is raunchy—really! Now
realize this—bzzzzzzz! Really.

Gee, I can see right now I'm going to be zapping trash all
day. Hello there! I'll bet your name is Sammy SlideRule!
Well, tell me, Sammy; what's the square root of 2232.16 to
the twenty-first power divided by 3 million? You have two
seconds. Bzzzzzzzz. Sorry, pal, wrong answer—and away
you go!

Oh, it's a walking calculator! Sure looks like a piece of
trash to me! Bzzzzzzz! Have a nice day. Boy, this is really fun.
My plan seems to be working, too. As long as I keep zapping
trash, no one suspects that I'm anything other than a real janitor. This is great! I could have used one of these vaporizer
gizmos back home. Oh, well. That's all water under the
bucket . . . I mean bridge.

Huh? What's everyone talking about? It's Elmo *who's*
birthday? Whoops, I'd better not ask that out loud. I'm probably supposed to know who this Elmo character is. Maybe
he's the boss. Gee, they all just wished him a happy birthday.
It must be nice to have people—especially your coworkers—
wish you a happy birthday. No one ever wished me one.
Well, I take that back—a couple of people *sometimes* remembered, but never the entire office. I guess they thought janitors weren't supposed to have happy birthdays. I'd probably
get stuck cleaning up after my own birthday party anyway,
knowing those jerks on Orbital Station 4.

Oh, well. Time to get back to work. This darn place runs
in circles, and I don't even know where I'm going. How do
these dopes even find their way around this maze? I must
have zapped at least 50 buckets of trash by now, and yet I
don't seem to be getting anywhere. Oh, gee. Who's that ogre
in the picture on the wall? Boy, I'll bet he's won enough nerd

contests. If not, he should start entering them. What a geek.
Hey, fella, whoever the heck you are, if you'd take those
Coke bottles off your eyes, maybe you could see something.
Nyak-nyak-nyak.

Hey, a copy machine! Boy, I love playing with these
things. There was one in the break room on the *Arcada*. My
buddy Willie and I used to make copies of our wallet photos
and then cut the copies in half and put different ones together to make new faces. Heehaw, we used to have a bang-up
time doing that! . . . Only now, my buddy Willie . . . *(sniff)*
. . . he's gone . . . *(sniff)*. . . . Oh, gee. I can't think of that right
now. They'll get suspicious if they see the janitor crying.
Boy, though, that copy machine sure brings back memories
of my days on the *Arcada*. That was such a long time ago.

Say, while nobody's looking, I'll bet I could get a quick
photocopy of that nerd on the wall. It's at least an 8 × 10, but
I'm sure I could put it together with some other pictures I
have. Yeah, why not? No one's over here to watch. Besides,
they all have their long noses stuck in their ledgers. Boy, I
could make some really funny stuff with this clown's face!
Hee-hee! I'll need some recreation when I finish with this
mission, so why not?

OK, this looks simple enough to operate, but these little
pictures showing you what to do are the lousiest excuses for
directions I've ever seen. That doesn't look anything like
someone feeding in the paper by hand. And they call that a
paper tray? Oh, well. This has to be simple—so easy, even a
janitor could do it. OK, the picture's on the glass—I'd better
hurry—and I'll bet this green button makes it go. . . . It
works! Oh, boy. I really feel like a Sub-Rosa agent now. This
is fun being sneaky and in disguise at the same time!

Great, the copier works! Oooo, he's as frightful in the
photocopy as he is on the wall. Now to fold this one up and
put it in my pocket—and put the original back on the wall
where I found it. Ciao, chump. Oh and hey, lose the specs,
OK?

Whew, that was fun, and nobody saw me! I'd better get
away from here as soon as possible, though. Someone might
wonder why it's taking me so long to zap all the trash at the
copy machine. Oh, well. If they ask, I'll just tell them some
slob left a bunch of papers strewn all over the floor. I'm the
janitor, right? It's my *job* to clean up those kinds of things,
right?

Gee, I think I've been this way before. Who knows how many times I've passed these same nerds over and over again—they all look so much alike. Heck, the only way I can tell whether I've seen them before is by their trash cans! Oh, I see a few more full ones in that direction. Maybe this part of the maze will lead to a pot of gold or whatever it is I'm looking for in here. Darn it, that's the trouble. I don't know what I'm supposed to be looking for. Every desk looks the same. Every nerd looks the same. Even their charts and file cabinets look the same. This is never going to end!

Well, on second thought, maybe I spoke too soon. That looks like an office over there—must be where the big boss works. Whooo, I'd better be careful around him. He might have me fired! . . . Gee, he just *might* have me fired—for real! That wouldn't be any laughing matter at all if this boss person recognizes that I'm not the usual janitor . . . (gulp) . . . but I have to go in his office. It's my JOB, both as a fake janitor and as a secret agent. Oh, well. Here goes. I'll just grit my teeth and keep my fingers crossed and hope to high heaven he doesn't pay me any attention!

(Gasp!) I don't believe it! It's the creep in the picture! Why, of course, I should have known. Who else would have his picture hanging on the wall—like a monument raised in honor to a great boss saint? And who else would they be wishing a happy birthday to? So this is Elmo. What a name for a boss. I'm not sure I could take someone with a name like that too seriously. Gee, he's just as nerdy in person as he is in that picture, and little does he know I'm carrying an 8 × 10 of him around in my pocket.

Oooh, what's that on his desk? Is that a *keycard* I see? Hmmm, it looks like I'll be paying this little office another visit—if this guy ever gets off his duff long enough to leave it. I don't think it would be a good idea to take the keycard while he's sitting here in front of me. I suppose I could "accidentally" knock over the trash can to cause a scene and then sneak the card off his desk. But no, I wouldn't want to risk raising St. Elmo's ire—that wouldn't be a good thing to do right now.

Saaaay, I just thought of something. This Elmo character must be the one holding the Two Guys captive! After all, he's the boss! Geez, I'd really better be careful. If he finds out I'm not who I say I am, the Two Guys and I will be goners for sure! OK, Elmo Whatever-Your-Last-Name-Is, I'll zap your trash for now, but sooner or later I'm going to zap your face out of this office. And then I'm going to zap it off the face of this moon. And finally, I'm going to zap you right out of this universe! You can't hold people in jello and expect to get away with it!

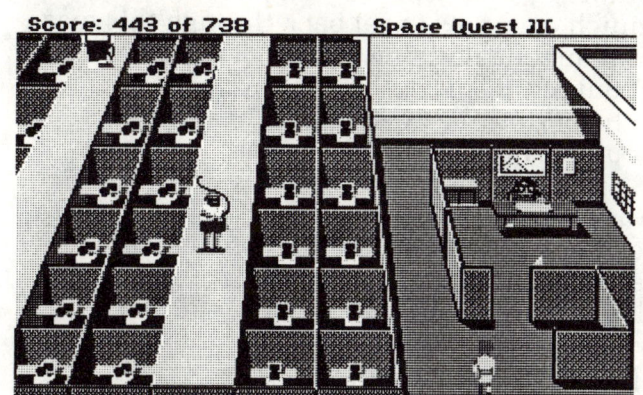

You made it through the maze, and now you've found the big cheese. Oh, and don't let those cracking whips bother you. They're just offered as "encouragement" to Elmo's legion of software developer "recruits."

Now that I've zapped St. Elmo's trash, I'd better leave. I hope *he* leaves soon, though. I need that keycard to get through that other door in the circular hallway. Maybe I can hang around here unnoticed for a little while, but I don't want to loiter around his office door. Hmmm, what's that door over there? Maybe *it* leads to the Two Guys. Well, since no one seems suspicious so far, I might as well try to take a look. Gee, look at those big oafs over there cracking their whips—talk about slave drivers! I'm glad I don't work for this scumbag place. Whoever named the company ScumSoft certainly didn't mince words.

Now, I wonder what's behind this door. Only one way to find out—I'll just open it, and . . . *(gulp)* . . . uh-oh. I do believe I recognize that spaceship over there. Yep, it looks like a recently refurbished battle cruiser of unknown origin, and it differs markedly from the those skull-faced fighter ships surrounding it . . . *(gulp)* . . . I've been discovered! But they couldn't know who I am—all they have is my ship. I can't take any chances, though. They'll find me out sooner or later. I have to rescue the Two Guys as soon as possible so we can get out of here. I can't blow this mission now! At least I know where my ship is. That wouldn't have been too cool to rescue the Two Guys only to find my ship missing when we escaped to the clearing. I hadn't figured out that escape part yet anyway.

Oh, look. St. Elmo left his office. I wonder if that keycard is still on his desk. I'll take a quick peek—I can't get to the Two Guys without it. There it is! Great, I'll just take that, thank you very much, and now to get back through this maze of cubicles without anyone noticing. Bunch of boring jerks—I can't wait to get out of this place. This is nerd city, and these guys are so busy running tapes on their calculators they probably wouldn't notice much more than an empty trash can.

I wonder if those guys back there at the computers were their software developers. But why are they so heavily guarded? Oh, wait a minute—judging by the looks of those skull fighters, I'll bet these people are . . . are . . . PIRATES! That's it! That's where I've heard the name *Pestulon* before. No wonder this planet was surrounded by a force field. They're the infamous Pirates of Pestulon! I'll bet the Two Guys infiltrated their sleazy operation, Elmo found out about it, and now they're holding them captive. Oh, darn. That means they've *got* to know my ship belongs to a Space Fleet Sub-Rosa agent.

Whew, I finally see the entrance to this room. I have to get out of here as quickly as possible. The Two Guys are already in imminent danger if Elmo and his thugs know about me. . . . Boy, I'm so glad to be out of that place. I hope I never see another plain brown desk or fluorescent light again! How do they work in there, anyway? The glare is awful!

Back in the Circular Hallway

OK, now I'm back on this merry-go-round again. Let's see—which way was it to that door with the keycard lock? I believe I passed it going north through this hallway, so now I need to go south. I just hope I don't bump my head this time—I could use a couple of aspirin right now between hitting the walls and the glare from all those fluorescent lights.

There's the door! I just hope this works. OK, I'll insert the keycard now, and hopefully, in a couple of seconds, I'll be in! Here goes . . . huh? What the heck is happening? Oh, no. They do facial scans here? Geez, I didn't know anything about that! I can't let them see my face! They'll never let me

186

in, and they'll know I'm an impostor! Think, Wilco; think!
What am I going to do? . . . The *photocopy*! Of course! That's
brilliant! I'll show them Elmo's face! Oh, where did I put that
piece of paper with his picture on it? It's in one of my pockets. . . . Here it is! I have to get this unfolded and then show it
before that scanner does its thing. . . . Whew, just in time! It's
scanning, and I don't hear any ruckus being raised. No mention of an impostor yet, either. Well, I'll be darned; it worked.
Give this man a cigar—I'm in!

The Two Guys' Room

Oh, my gosh. And there they are, frozen in slimy green jello.
What a horrible fate—and messy, too. I was right—this is the
room with the Two Guys. Look at them, those poor creatures—my colleagues. But how do I get to them? They're
stuck in jello on a platform in the middle of the room—and
there's no floor in between. "Hold on, Guys. I'm Roger Wilco,
special agent with the Sub-Rosa Service, just like you. I'll
have you out in a jiff!"

Now what do I do? They're expecting me to act like I
know what I'm doing. They're probably watching everything
I do, making mental notes and stuff like that. One of these
days, I'll be trained to make mental notes. "Just hold on a little longer, Guys. I didn't come all this way only to be stopped
by a big hole in the floor, ha-ha." Gee, I'm starting to sweat.
This is just a little embarrassing. What if they go back and
tell everyone it took me five minutes to figure out how to
cross the hole in the floor before I could rescue them? There
has to be a way to bridge that opening.

Wait a minute. What's that button by the door? It could
be an inside door button—but it's worth a try. There must be
something around here that opens up a walkway to that center platform, so maybe that button will do it. Besides, I don't
see anything else to try. OK, here goes. . . . All right, it
worked!

"I told you I'd be there in a jiff. Just hold on, Guys. I
think I have a way to get you out of that jello, too. Don't worry; I don't think this will hurt. I'm going to zap that jello with
my Mr. Garbage. It's supposed to work on food, too, so just
hold tight. OK, here goes. . . . Hey, it worked! I told you I'd
have you out in a moment.

"By the way, how do you do? Again, I'm Roger Wilco, one of your fellow Sub-Rosa agents. Sorry for the delay in getting here, but I think the Space Fleet wanted to test my abili— Huh? What do you mean what's going on? What's with you Guys, anyway? What did they do—turn your *brains* into jello? . . . I said I'm your fellow Sub-Rosa agent, and . . . *what*? You're *not* with the Sub-Rosa? You mean you're not Company men?

"Uh, look, Guys. I know you've been under cover for quite a while now, but you don't have to pretend anymore, OK? I'm one of you, get it? . . . Give me a break. Didn't you hear what I just said? I think St. Elmo *did* turn your brains into jello. *Read my lips.* I AM A FELLOW SUB-ROSA AGENT. Like I said, you don't have to worry about blowing your covers around me because I already know who you are.

"What do you mean what covers? Your Sub-Rosa cover, you idiot. What other kind of cover is there? I know your cover is Two Guys . . . huh? I don't get it. What exactly are you saying? You're telling me you aren't with the Sub-Rosa service? . . . But I thought. . . . Well, just who the heck are you then?

"You're WHAT? I don't believe this. You're kidding, right? This is a joke, isn't it? . . . Do you mean to tell me I risked life and limb battling scorpazoids, Arnoid the Terminator, a red-hot planet that just about blew up in my face, a maze of cubicles full of nerds in starched shirts, and countless other things to rescue a couple of guys from Andromeda who write software for a living? All this time I thought I was on a secret mission, and I've just been on a wild goose chase, taking all these dangerous risks for NOTHING?

"Oh, gee. I'm not believing this. It makes me faint to think of everything I've been through, and now to find out it was all for . . . for . . . Astro Chicken people? I mean, it's a good game and all, but let's face it—this isn't the same as rescuing fellow undercover agents. Not only that, this all means that I never was—and still am not—a James Bond. I was never recruited secretly by the Sub-Rosa Service or any other Space Fleet organization. This has all been for NOTHING! . . . Oh, quit asking me so many questions. How the heck should I know? I'm just a janitor from Xenon—and I always will be.

"Huh? What's that? Oh, no. I knew it! They found me . . . us. I knew they'd figure it out sooner or later. Oh, go suck your thumb, Elmo, you big fat baby. Who cares what you think, anyway?" What in the world is he blabbing about? What arena? What showdown?

Nuke'm Duke'm Arena

Hey, I don't believe it! It's the Nuke'm Duke'm Robots! All right! I used to play this game all the time when I was a kid! Oh, boy. I'll bet Elmo thinks he's got this one in the bag. Well, just wait 'til he battles it out with the Nuke'm Duke'm king!

Rock'm Sock'ms were nothing compared to Elmo's Nuke'm Duke'ms.

"Hey, you Two Guys, don't worry about a thing!" Boy, they're the biggest couple of scaredy-cats I've ever seen. I'm sure they don't think I can win this round, and obviously Elmo doesn't think so, either. These robots are so huge. I remember the first time I set foot in one as a child. I felt like I was climbing a ten-story building. The only difference then was that it was all for fun. This guy is playing for keeps! Gee, I sure hope I remember all my good punches!

I know one secret to winning this game—every good punch means more energy for me. As long as I can go in and punch his lights out without stopping, I can probably win the whole thing! And as long as he keeps punching at me and missing, he'll use up his energy so fast it will only take one good swipe to knock him out. I'll make him think I've never even seen a Nuke'm Duke'm Robot before. That should really throw him off track.

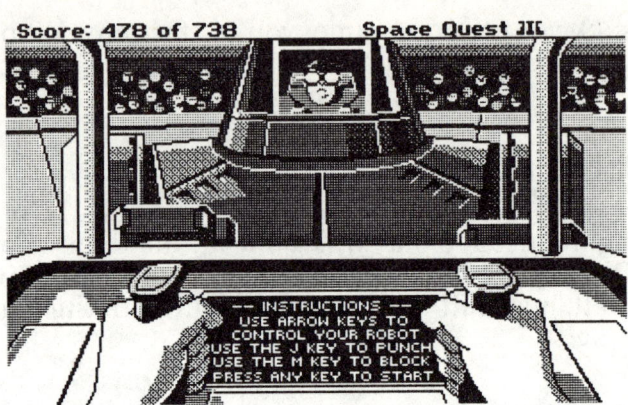

Score: 478 of 738   Space Quest III

-- INSTRUCTIONS --
USE THE ARROW KEYS TO
CONTROL YOUR ROBOT
USE THE J KEY TO PUNCH
USE THE M KEY TO BLOCK
PRESS ANY KEY TO START

Elmo's face is all the inspiration you need to knock his block off.

"Hey, wait just a minute, *Elmer*. No fair! I didn't have time to read the rules. How do I know you're not going to cheat? Oh, well, excuuuuse me! I thought it was Elmer. You look like an Elmer—Fudd, to be exact." Nerd. He's too old to be playing these silly games, anyway. That's right; just keep thinking you have the edge. We'll see about that; won't we, Elmo?

Here we go. . . . "Take THAT, Mughead . . . and THAT and THAT and THAT!" Oops, he threw me a good one! "I'll get you back for that one, you ScumSoft scumbag! You'll be so nuked and duked when I'm through with you you'll want to puke! You got that, ELMO? Here, take another knuckle sandwich, blockhead!" Oh, boy. I can see I've taken him off guard. That jerk didn't think I had a prayer out here. I'm showing him! Just a couple more good punches, and then I'll go for the knockout!

"Hey, Elmo, look! What's that over there? Aha! Fooled you, sucker. Take THAT and THAT! . . . Huh? What's what over where? OUCH!" Ooooo, that dunce got me back. How did I fall for my own trick? That was pretty stupid, but I won't do it again. "Take another NUKE, Elmo. How about another DUKE, Elmo? Are you ready yet to PUKE, Elmo? . . . Elmo? Hey, where'd he go? What? I did it! I did it! I punched his living daylights out!"

Uh-oh, now where are the Two Guys? "Hey, you Guys, come on! Let's get the heck out of here—on the double!"

In the Battle Cruiser

"You Guys just have a seat there while I get things going. Look, would you just calm down? You're acting like idiots, and stop asking so many questions, will you? You're not helping things!" OK, engines, radar, hmmm—I'd better engage the weapons system this time. Those pirate ships are likely to be right on our tail. "OK, folks, we're off—and not a moment too soon. Uh-oh, what's that coming up behind me? Hold on tight, Guys. I have a feeling we're in for the battle of our lives—literally! And I don't know how much this little cruiser can take. I found it on a junk ship, you know.

"Yeah, that's right, a junk ship. But don't worry; she's done pretty well for me so far! I got to Pestulon, didn't I? Now just calm down!" OK, back shields are on—oh, gee. Would you get a load of those things! They're even creepier looking when they're flying through the black of space! I've got to zero in on him . . . and . . . FIRE! "I got him! I got him! Did you see that? Did you? Well, did you?" Uh-oh, here comes one from the front! Better switch to my front shields el pronto.

Oh, they got me! Darn it! "Hold on, Guys. Just keep good thoughts. I don't know how long the front shields will hold up, but we can't take too many more hits like that!" They're much harder to take coming from the front. Uh-oh, radar is picking up one from the back again. . . . I almost have him . . . almost have him . . . FIRE! "Oh, boy. I did it again! Did you see? Well, did you? Am I a great combat pilot or what?" Oh, no. They keep switching sides. Here comes one from the front again. . . .

After the Battle

"Whew, we made it! I don't think you Two Guys realize just how close that was. We almost completely lost our front shields! If I hadn't made all those hits, we probably wouldn't be here right now. I don't think we could have taken another one from the front.

"Let's just get the heck out of here, OK? Oh, no. Something's wrong with the light speed control. Darn it! We can't go anywhere without that . . . huh? You do? Well, then by all means, please try to fix it!

"Gee, thanks, Guy. That was really swell of you to fix the light speed. Yeah, I know we're in this together. Well, I have to admit I was pretty ticked off when I found out you were really just Two Guys from Andromeda and not envoys like I thought. I just *knew* I had been recruited by Space Fleet.

"It's kind of a long story, how I got here and all, but what the heck—we have plenty of time. You see, it all started when I realized at the ripe old age of five that I wanted to be a janitor. I wanted to help clean up this universe, and. . . ."

End of Game

Sorry, this part's a secret! You'll have to make it through the battle with the skull fighters to find out what happens next. Good luck, though! See you in *Space Quest IV!*

What sort of strange new world lies ahead?

What ingrates, those Two Guys. They made a bee-line to get off the ship when you landed, and now they accept the first job offer that comes along. Guess you weren't kidding when you said it was a long story.

Game Points Earned

Action	Points
Entering the ScumSoft building	25
Finding the Two Guys	20
Getting the keycard	5
Photocopying Elmo's picture	5
Putting on the coveralls	5
Vaporizing the jello	10
Winning the Nuke'm Duke'm Robots game	100
Winning the space battle	100

Object Locations

Object	Location
Coveralls	Janitor's closet at ScumSoft
Keycard	Accounting maze, on Elmo's desk
Photocopy of picture	Made with the copier
Picture of Elmo	Accounting maze, near the copier
Trash vaporizer	Pocket of coveralls

Chapter 8
Roger Wilco and the Time Rippers: Lost in Space Quest

Gee whiz, Wilco! It couldn't happen to a dumb . . . er, nicer guy. By now you must feel like you're in some sort of crazy adventure series—you know, where they keep churning out sequel after sequel. Well you might as well get used to it, Roger. This one's not over 'til it's. . . . Well, who knows when. Why should you care anyway? Apparently this hero stuff is your true life's calling, despite your janitorial inclinations of the past.

Your biggest problem right now, unfortunately, is the same *big* problem you encountered before, only this time, it's massive. That's right—Sludge Vohaul is back in town. Somehow old Fatback managed to survive—in essence, anyway—both the destruction of his asteroid and the termination of his life support system. Old Sludge always did have a short—albeit fat—fuse, and like the elephant he is, he never forgets. In other words, he's more than just a little ticked off by the shenanigans you pulled on that asteroid.

You may think you're Roger Wilco—space man, hero, ex-janitor, and sometimes tap-dancer—but he still remembers how silly you looked hanging onto that plunger over the acid pool. To him, you'll always be just an annoying, pea-brained little bug in his computer—but one he'd desperately like to squash.

Unfortunately for Xenon—and the entire universe for eons to come—Vohaul has been utterly single-minded in his determination to rid you from the universe. He's pulled out all the stops and left no stone unturned. And that's your second biggest problem. Now that the secret of time travel has been unraveled, you don't know *where* to go—back to the future, forward to the past, or stand still in time.

Truth be known, it really doesn't make any difference where you go because you'll never get away from Vohaul. He's like your Bank of Xenon gold card—he's everywhere you are. And that's your third biggest problem—you never seem to know where you are. Until you find a way to put Vohaul out of business permanently, it looks like you're destined to remain . . . *lost in Space Quest!*

Total Possible Game Points: 340

The Bar on Magmetheus

". . . And then I grabbed that blame Sarien by his tacky red collar, knocked his pulseray to the floor with my free hand, and then threw him all the way down the hall. The Deltaur's floor was so slick—well they did use a pretty good wax, you know—that jerk slid all the way down the hall on his as. . . .

"Huh? Yeah, another round of cocktails for my fine, fleshy friends here. Say, have you boys ever been to Kerona? Well don't bother. That's the worst beer I've ever put in my mouth. And speaking of places to avoid—forget about Earth. You know those old expressions—Where on Earth? How on Earth? What on Earth?—all those old sayings you hear all the time? Well, by golly, now I know how they all originated. 'Where on Earth did you drag up that thing?' makes a lot of sense to me now. Believe me, you don't want to bother with that black hole of a place.

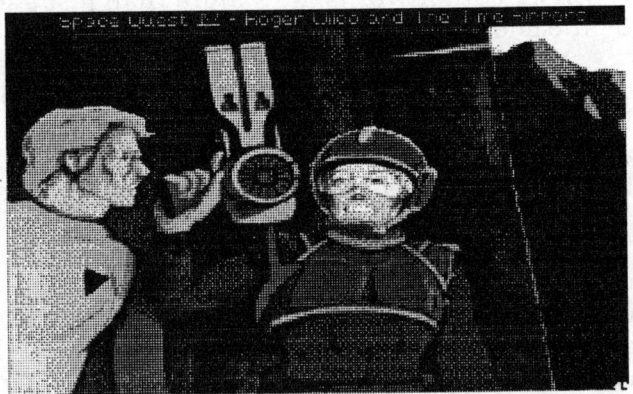

Who IS that masked man?

"Anyway, as I was saying, that darn Sarien slid so fast. . . . Huh? Yeah, I'm Roger Wilco. Who wants to know? Oh OK, sure. I didn't do anything wrong, did I?"

Gee, these guys sure are armed heavily for traffic cops. "You're what?! *Sequel* police?!" Oh no. I'm not sure what that is, but I don't like the sound of it. "Ruler of Xenon? *What* ruler of Xenon? Xenon's a democracy, you nincompoop—at least it was the last time I was there. Is this some sort of joke or something? . . . (*gasp*) . . . VOHAUL!

"Oh no! There's no way. This can't be. I don't believe it. There's no way that rotten egg can still be alive! I got rid of you a long time ago, you sewer slug. I don't believe any of this. You're not real. You're just some stupid hologram someone's rigged up to scare me. Well I've got news for that sucker. Holograms are old technology. They're on every magazine cover around, and I'm not afraid of them."

Huh? What's happening? Who are those guys, and why are they dressed like that? What are those weapons they're carrying? Why do I always ask myself questions I can't answer?

Look, they've knocked out the Sequel Police! "Come with you?! Why the heck should I. . . . Y-y-y-yikes! W-w-wait for me! I'm coming! I'm coming!

"Whew! That was close! Hey, what the. . . . Huh? Who *are* you? What do you mean they have a *bead* on our coordinates? What the heck kind of terminology is that? Who do you think I am, Noah Webster? What's a *time rip*? Look mister, I have a lot of questions. . . . All right, all right—I'm going. Don't push. I mean it! Don't push! I can jump by myself, but I want some answers when I get . . . baaaaaack!"

Space Quest XII: The Planet Xenon

Current Goal: Escape safely from the planet after hearing the story of Xenon's fate.
Total Points: 63

The Planet Xenon

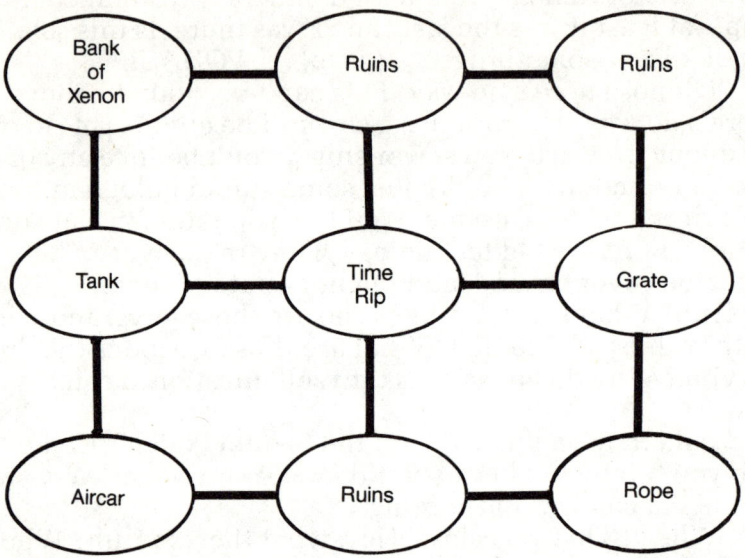

Inventory

buckazoids
Energizer bunny
glass jar with slime
mortar
PocketPal minicomputer
string

South Xenon Ruins

I don't know what just happened, but one thing's for sure—if I ever get back home, all those rainbow plaques on my wall are history. And I'm taking that rainbow air freshener out of my aircar, too. . . .

Hey wait a minute, where the heck am I anyway? This place is a mess! Don't these people believe in curbside garbage collection? What is all this stuff anyway? Gee . . . *(gulp)* . . . it looks like ruins. No wonder there's no cleaning detail. I've never seen such devastation and destruction in my life—at least not this close. And gosh, if I didn't know better, I'd think this was . . . was . . . Xenon! But this *can't* be Xenon! There's nothing here. And that monstrosity out there—what in the world *is* that thing? At least it's alive—there are lights in the windows—which is more than I can say for the rest of this place.

Boy oh boy, though, if this doesn't all look so familiar to me. Honestly, if I didn't know better, I would swear I'm standing in the very spot where I stood when they handed me that golden mop award for defeating the Sariens. Still, there's no way this could be Xenon.

Maybe in a couple of hundred years . . . *(gasp)*. . . A couple of hundred years?! Oh no! What was it that guy with the funky hairdo said about saving Xenon? And what did he call that thing I passed through? He said it was a *time rip!* That means. . . . Oh no, I don't believe it! This IS Xenon! I'm in the *future!* I passed through a *time rip!* So what if it's really supposed to be called a time *tunnel*—it's the same thing!

You'd like nothing better than to lasso that wacky wabbit. Hey, now there's an idea!

Ahhhh! What's that noise? . . . Oh, whew! I don't believe it! It's the Energizer bunny, and it's still going! Man, I should have stocked up on those batteries when I had the chance. What with inflation and all, and now all this destruction, I'll bet the price of batteries will go sky high.

That certainly is a spastic little bunny. I can't even get near the thing without it turning around and marching off. Gosh, that brings back so many memories. I used to love those commercials. I know one thing—if I ever make it back to my own time, I'm going to invest in Energizer stock. Evidently, it's the only thing that lasts.

Gosh, I can't believe what's happened here. Surely there must be some Xenonites left. Were they *all* destroyed? And who's keeping that . . . that . . . building thing out there illuminated? Oh, wait a minute! Dollars to doughnuts it's VO-HAUL! But of course! Why else would that Time Ripper guy tell me I was Xenon's only hope?

Who else would do this sort of thing anyway? I wasted the last of the Sariens a long time ago, and that Elmo character—he was too involved in his accounting books and pirated software to exact this kind of revenge. No, I would expect ol' Elmo to sell me a phony hint book or something if he wanted to get back at me.

Now I know I *must* take a closer look around. Besides, what choice do I have? I have no place to go and no way to get there. First I'll go north as far as I can. Maybe I can get a better look at that—whatever it is out there.

Far North Ruins

Hmmm, it looks like I've reached the edge of the city—or what's left of it. I still can't tell what that thing is—some sort of building, I think. What an odd shape, and what's all that lightning-looking stuff coming out of it? It must be some sort of power generator, but power to what? *It's* the only thing with any life to it, so maybe it keeps itself alive. Everything else here is dead—except that Energizer bunny. Here he comes again—and there he goes again. Strange little toy, that one.

Bank of Xenon (Northwest)

Say, what's that building over there to my west? It looks
like—it is! It's the Bank of Xenon! I can't believe it. Every
dime I owned was deposited through those revolving doors,
and now look at it. Some savings plan that turned out to be,
and to think I trusted my personal banker. Shoot, I never did
believe all that junk about the banks being insured by the
Xenon government. Now look at it. There IS no Xenon government, and the only buckazoids to my name are the few
jingling around in my pocket.

Hey, I wonder if there might be a few buckazoids lying
around on the floor inside. Now, if I could only get this . . .
blasted . . . door open! Darn it! I can't get inside that bank.
Those doors are stuck for good, and I sure don't have anything that would pry them open. Oh well, I might as well
keep looking around.

Abandoned Tank Area (West-Central)

Hey, what's that?! Oh boy, it's a tank! Wow, a real, genuine
combat tank used in a war! This is great! I don't believe it!
I've never seen one of these for real! Wow, it's incredible!

Oh boy, I've always wanted to get inside a combat tank
and pretend I'm right in the middle of a battle! Yeah, that's it!
I'm in a ground war. The fly-boys have done their duties, and
now it's up to us tankers to move 'em in and clean 'em out!
Rat-ta-ta! Brrrrrrat-ta-ta-ta-ta! Boom! Ch-ch-ch-ch! Boom!
Ch-ch-ch-ch! Brrrrrrat-ta-ta-ta-ta-ta-ta!

Boy, this is fun! I wonder if there's anything left in. . . .
Hey, look at what I found! It looks like leftover mortar! Wow,
what a great souvenir! I'm taking this with me! Besides, it's
probably dead—just like everything else around here—everything except that . . . that . . . *person* headed straight toward me!

Oh my gosh, it really is a person! It's someone alive! He
looks pitiful—he must be injured. But I'm a janitor, not a
doctor. Oh what the heck, maybe I can help anyway. Gosh,
he looks like he could be somebody's grandfather, poor man.
There's no telling what he's been through. Maybe we can
find a way out of this ghastly place together.

Uh, he must not have very good vision. He's headed
straight toward me, and yet he isn't really *looking* at me. Gee,

I can't imagine why he wouldn't be looking at me, considering that his eyelids have been permanently pulled back by two—WIRES?! And look at his teeth! I can see all four of them from here. The rest must have rotted out—not that it's his fault or anything.

We're talking looks that kill—literally. Try to keep your distance from this guy.

Oh no, that smell! Oh, it's horrid! He isn't even close to me and already I could faint dead away from that stench. Honestly, if I didn't see that . . . person . . . walking with my own eyes, I would think he was dead. He smells dead.

On second thought, I don't think I want to hang around in time for this fellow to get too chummy. He's dragging his leg and all, and, well, maybe he just isn't in a great mood. Besides, the guy really doesn't look quite right. Maybe that's selfish of me, but I can't be too trusting . . . *(gasp)* . . . I'll bet he's working for the enemy!

I'll bet he's some sort of half-human, half-machine thing assigned to patrol these streets! In fact, Vohaul probably has him looking for me! Now I know I have to get away before that guy—or whatever he is—sees me. I better go back the way I came. Adieu to you, Led-foot.

Northeast Street

At least that creepy old man moves slowly—guess he can't help it, dragging that foot around and all. I could outrun that dude any time, thank goodness. Gee whiz, even a biped could outrun that geezer. I just wonder what would happen if we actually did come face to face. I believe I'll pass on finding out, just the same.

Oh, there's the little bunny again. Boy, those batteries really do just keep on going and going. I'd like to be the guy who got a patent on those things. Well, there isn't much to see around here. I can't seem to get in any of these buildings. I suppose the doors were all boarded up and welded together on purpose. How could anyone survive around here? I think I'll see what's south of here.

Sewer Grates (East-Central)

Oh, now *this* is interesting—sewer grates. Boy, I'd recognize those things anywhere, and especially here on Xenon. I wonder if the sewer system is still intact. Hmmm, I'm glad I found these grates. If those ducts are still open below, this sewer system could prove to be an excellent hiding place. I'll bet Gramps couldn't drag his leg down these manholes.

I'll remember the location of these grates. Maybe I'll even do some exploring later, but first I want to take another look at that tank. One thing's for sure, I'm not going back there the way I came. I expect to see that geezer dragging around the corner any minute now.

Southeast Street

Say, what's that—a piece of string? Eeew, I'll bet it came off that geezer's pants. Yuck. Well, it does look to be pretty much intact. Maybe I'll keep it. . . . Who knows when a string might come in handy, and I've learned from experience to take everything I can find.

Now to get back to that combat tank. I wonder if there are any more old weapons hiding in there. Gee, the only thing about carrying this mortar around is that I don't know whether or not it's stable. I mean, I may need it at some point, but what if it blows my head off when I'm least expecting it? . . . (gulp). . . On second thought, maybe I'll just put the mortar back where I found it. It's kind of heavy to be carrying around anyway. I could always come back and get it in an emergency, couldn't I?

There's that darn bunny again. I'd like to snatch that little fellow up and learn the secret to his Energizer battery.

But darn it, I'm an ex-janitor, not a chemist. How would I know what to do with the insides of a battery? Might as well nix that idea.

Uh oh, here comes Mr. Universe again. Where's the fire, buddy? Geez, I think I'll just hide behind one of these pillars this time. He never seems to change directions. I'll just let him get a few paces in front of me. Of course, that could take hours—literally—but then at least I wouldn't have to look over my shoulder every five minutes.

OK, any day now. Let's take all the time in the world we need to drag our lame, led foot around and make our way through these bustling streets. "Hey Good-lookin', need some help? How about some mortar for your face? It'll do wonders for those crows feet. And I promise—it'll be a real blast. Nyak-nyak-nyak." Ooops! He's coming over here. Oh golly, I hope I didn't say those things too loud. Surely that walking corpse can't hear any better than he sees. Whew! Thank goodness! OK, Wilco, cut out the smug stuff. I'm not out of the water—er, ruins—yet.

Back to the South Street

Well, I believe I've passed this way before. If I'm not mistaken, this is where I landed when I fell out of that time rip. Just a couple more city blocks to cover and I'll have made my way full circle around what's left of Xenon's main city.

It's still so hard to accept that this is Xenon—my Xenon—somewhere in the future. I guess this doesn't have to be the future, though, not if I can find a way to get rid of Vohaul forever. What a laugh—I don't even know where he is. I don't even know *when* he is. For that matter, I don't even know *when* I am. I know where I am, but that's about it, and that's not helping a great deal. Time to mosey on another block or two.

Abandoned Aircar (Southwest)

Hey now, what's this? It looks like an aircar! Wow, I wonder if it's still working! Oh gee, would you look at the size of the hole in this thing. There's no way this baby's going anywhere, darn it. If only I could fix it the way I did that battle cruiser on the junk freighter.

Boy, that seems like lightyears ago. It probably was!

Well, at least I can explore a little and see if anything important was left behind in this aircar. . . . Nothing under the seats. . . . Aha! The glove compartment! Why didn't I think of that to start with? . . . There's that darn bunny again. That cwazy wabbit is weally starting to bug me. Why does it have to look so cheerful when everything else around here screams gloom and doom?

Hey, what's this? Wow, it looks like some sort of mini-computer! Well I'll be—it's a PocketPal. You mean they're still making these things? Gosh, I must not be too far into the future—PocketPals have been around forever. I'm definitely taking this with me. I wonder if it works. . . . Let's see. . . . Shoot, there's no battery in it. Darn it. Oh, wait a minute! The Energizer bunny! I'll get the battery from the bunny!

In fact, it sounds like he's coming back here right now. . . . No, wait a minute. That doesn't sound like a drum beating. It's more like a humming noise, and it's getting closer and closer! I don't know what the heck I'm hearing, but I'm not hanging around to find out! Y-y-y-yikes! Where can I hide? That overhang—I better make a dash for it! What if it's one of Vohaul's cronies?

Huh? Great balls of fire, what in the world was THAT!? Some kind of mechanical clicking thing—a droid? But what would a droid be doing around here? That's the first I've seen or heard anything like that since I've been here. Gosh, I wonder what it would do if I hadn't hidden?

Actually, I don't think I want to hang around long enough to find out. That combat tank's just up ahead. More and more now I think I should put this mortar back in the tank—especially if my next move is to climb down one of those manholes beneath the grates. If this thing is unstable—and it very well may be—squeezing down into one of those manholes could mean curtains for me!

Back to the Abandoned Tank

OK, back into the tank you go, Mr. Mortar. Now, to make my way over to the grates. Gee, I sure hope that droid isn't hovering around over there. I have a feeling I'm as good as dead if he spots me!

Back to the Sewer Grates

Whew! I made it back to the grates without being detected by either the geezer or the droid. Oh no! What's that noise? . . . Oh, thank goodness! It's that wacky wabbit again. Come here, you crazy pink bunny. . . . Darn it! I can never get close enough to that guy. How does he do it every time?

Saaay, I have an idea. I could use that piece of string I found to set a trap. Then, when that little fellow comes marching through with his big bass drum—the way he does every five seconds—I'll snatch him up, and the battery will be mine! He does seem to be intelligent for a battery-operated rabbit, though. Gosh, even toys are sophisticated these days.

I'll just hide behind this beam. Hopefully, he won't be able to detect me. Oh boy, here he comes! Come on, just a little closer. . . . Got him! I got him! I got him! Nah nah nah nah, I outsmarted the Energizer bunny! Let's see who keeps on going and going now! . . . Uh oh, there's that humming noise! The droid must be making his rounds again. Y-y-y-yikes! It's now or never for exploring the sewer system, and I vote to explore it! The cityscape was really beginning to bore me anyway. . . .

Xenon Sewer System

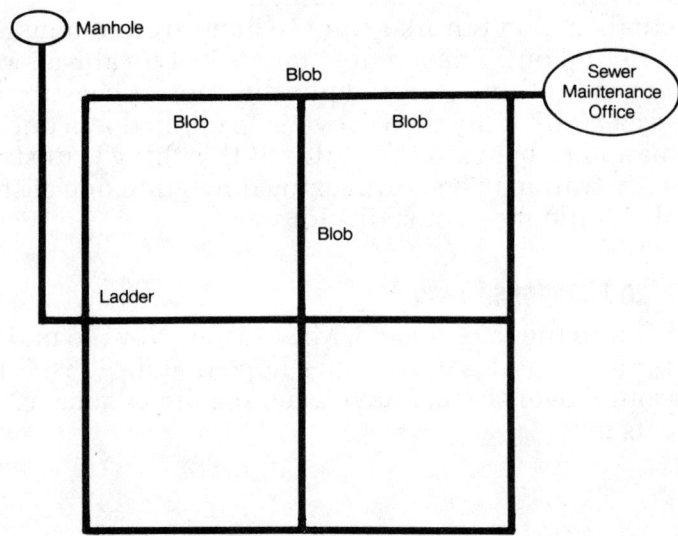

Sewer Maintenance Office

Oh wow, I recognize this place! This is Xenon's main sewer-system maintenance office—and it's still intact! Wow, this is unbelievable. Surely there's something in here I can use—something that will help me find a way off this planet. Boy, I'll bet this is the only place on Xenon Vohaul didn't destroy, and I'm sure that's because he never knew it was here.

Help him Obi-Wan . . . er . . . Professor Loydd. You're his only hope!

Let's see, there's bound to be something in this desk . . . huh? This is stupid! I should have known—cheapo desk. It doesn't even have any drawers. Well darn it, obviously *somebody's* been in here or else there would be papers or something here—anything! There's nothing. Just an empty jar and a desk blotter.

The jar I'll keep—I might find a use for it later—but the desk blotter stays here. I'm not about to lug that useless thing around. Hey, wait a minute! I'll just bet someone hid a few secret documents under the desk blotter! Of course, that's the way it's always done!

What? Nothing's under there except a button. Oh well, there's only one way to find out what it's for. . . . Here goes. I'll just push the button and. . . . Holy star wars! It's Obi-Wan . . . er . . . I mean, it's an old, bearded man wearing a long robe! His name is Professor Loydd! He's the lead designer of the Xenon Super Computer project! . . . The *what?!*

Xenon never had any Super Computer—but of course, this is after my time. Wow, he says it's the ultimate in artificial intelligence. This is unbelievable. . . . The Super Computer was built to enhance life on Xenon until some piece of infected trash called *Leisure Suit Larry* was salvaged and then uploaded for analysis. Gee, it sounds like the computer wasn't the only thing infected—his colleagues apparently lost every shred of self-control and decency over this Larry thing.

The Super Computer became infected with a virus and stopped all operation until . . . HUH! . . . the words *Wilco Must Pay!* appeared on screen! Oh my gosh! It was Vohaul who infected the computer! I'll bet he set that *Leisure Suit Larry* garbage out as bait—and it worked! But how? How could Vohaul have survived all those years?

Xenon's computer waged war against its own people? . . . They were *modified*? . . . (gasp). . . . So that's what that old geezer is all about. He's a . . . changeling—only an old one! He's a . . . a . . . *cyborg*! Ahhhhhhhhh! And to think I almost came face to face with that thing! He would have turned me in. I know he would. That's what the Professor just said. He said they infiltrated the resistance movement and found almost all of their hiding places . . . and they're still roaming the streets!

Oh gee, this is almost too much to handle. I think I'm going to faint. Oh, but I can't; there's still more. Oh, now I see—it was the machine that unraveled the mystery of time, not Vohaul. I knew that bucket of slime was too stupid to figure out something like that. But still, how did he survive all those years—unless *he* was the virus?! Oh, gag me with a spoon—that figures. That filthy blob of mold—that's how he did it! So those men—the Time Rippers—they were from Xenon sometime in the future—*this* future! And since they brought me here, I'm the planet's only hope? Great, just great.

One thing's for sure. Sludge Vohaul is definitely a virus—he always has been. He's been a giant germ just waiting to spread for as long as I can remember. If I have anything to do with it, that's one virus who's run his deadly course for the last time.

Well, this certainly turned out to be an enlightening room. Too bad I can't hang around. I have to find my way to some sort of time portal or tunnel—whatever it is they use for time travel. I can't go back out to the Xenon surface, though, not after what I've just learned. I would be so scared if I saw that cyborg again, I'd probably freeze in my tracks even though I know I can outrun him. I'll have to stay below the streets until I figure out what to do. At least I know my way around here pretty well. I know exactly where that door leads—straight into the sewer conduit. I spent many an hour in these tunnels at one point in my life—but that was *years* ago!

Sewer Tunnel

Well, certainly nothing has changed about the old sewer conduit. Let's see, I believe if I go south from here, I can go all the way around to the other side, and there should be a ladder leading to the street above. At least I could steal a peek at what's going on up there—if anything, that is.

Huh? What's that noise? I don't remember hearing anything like that before. It's almost like a gurgle. . . . Ahhhh! What is that?! Eeew, yuck! Green slime! It almost looks alive the way it came through that vent and now appears to be following me! Oh, no way! There was never anything like that down here before.

Maybe I should collect a little sample of it—if I can stand the smell. I'll just scoop up a little in this jar and worry about analyzing it later. Saaay, this stuff is practically bubbling. What is it, anyway, some sort of acid?

Y-y-y-yikes! Get away from me! Ah! I'm not letting any bubbling blob of acid slime touch me, noooo sir! I came close to having my skin reconstituted by this stuff on Vohaul's asteroid. Vohaul—that's it! This is some more of his crap! Oh, I've about had it with that freak.

Y-y-y-yikes! I think this slime blob is alive—it's following me everywhere I go! The next thing I know, it'll be pouring though the ceiling. I've got to get over to the other side fast! I just pray there's a ladder leading up to a manhole like I remember. If not, I'm in deep . . . uh . . . slime!

Whew! There it is just up ahead. I knew there was a ladder in here somewhere. I'm glad I have such a good memory for detail—this kind of detail anyway. Heck, I remember the layout of every sewer system I've ever serviced. Now to climb up this ladder and see what kind of view I can get from the street. With my luck, that cyborg will be the welcoming committee.

Patrol Ship Site

Saaay, something IS going on out here. Is that a patrol ship I see? It is! Oh no, it's the dreaded Sequel Police! They're looking for me; I just know they are! At least no one saw my head pop out of this manhole.

Hey, now that they've scattered, maybe this is my big chance. If I could make it to that patrol ship without being seen, I'll bet I could figure out how to fly it and then get the heck out of here. It's worth a try. Besides, there's no alternative. Xenon's surface is just one big dead end. This is my only chance—so here goes!

Time is of the essence here. As soon as those SP guys scatter, lose the tin derby and make a beeline due east!

I don't see anyone! Wait a minute—the Sequel Police! There's one poking around the old Bank of Xenon. . . . Whew! He didn't see me! . . . Hey! This ship must be on remote or something. It's flying itself—and we're headed straight for that lighted building!

Game Points Earned

Action	Points
Capturing the rabbit	10
Climbing out of the manhole	3
Escaping in the patrol ship	5
Getting the jar	5
Getting the ordnance	25
Getting the PocketPal	5
Getting the rope	5
Going underground	5
Hiding from the death droid	5
Returning the ordnance	−20
Scooping up the slime	5
Viewing the hologram	10

Object Locations

Object	Location
Buckazoids	In your pocket
Battery	In the bunny
Energizer bunny	Xenon surface
Glass jar	Sewer maintenance office
Green slime	Sewer conduit
Mortar	Abandoned tank
PocketPal minicomputer	Abandoned aircar
String	Southeast ruins

Xenon Super Computer Complex: Landing Bay

Mission: Safely steal away in one of the time pods.
Total Points: 10

Xenon Super Computer Landing Bay

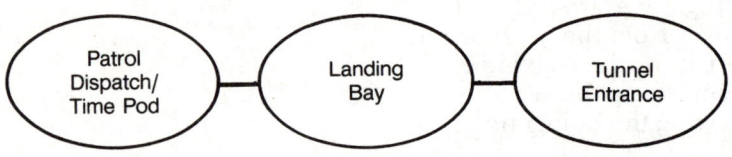

Super Computer Landing Bay

Hey, we're landing. . . . Now that was pretty smooth. I have
to admit, I could get used to this kind of space travel. The
ship just flies and lands itself. None of that crashing-into-
the-desert junk like I experienced back on Kerona.

Well, now that I'm here. . . . Saaay, this is one of those
lighted windows I saw from Xenon's surface. That means I'm
either in the middle window and there's a room on either
side of me, or I'm in the left window and there are two rooms
to the right of me, or I'm in the right window and there are
two windows to the left of me. Well, duh, this shouldn't be
too difficult to figure out.

Boy, just look at Xenon—I can't believe it. I can't believe this is the planet I once called home . . . *(sniff)*. . . . Whatever happened to Fly-Boy? . . . *(sniff)* . . . and what about Gladys? Gosh, she was so swell, and I never even had a chance with her. I can't bear to imagine their fates, especially after hearing Professor Loydd's message. They probably turned Fly-Boy into one of those humming death droids. And Gladys—oh no—they couldn't have! They wouldn't have! What if they made her into one of those cyborgs and she's still roaming around the streets of some city on Xenon? Ack! I can't bear the thought. I won't think about any of that right now. I'll think about that later. The only thing that matters for the moment is getting rid of that evil plague, Sludge Vohaul.

Stand around killing time (like those two flunkies) and it will eventually kill you.

By the way, where exactly am I? I mean, I wonder if those three windows are part of the Super Computer structure Professor Loydd talked about in his hologram message. Yes, that's the only possibility—this must be it. As far as I can see from here, it's definitely super in size.

Now, where should I go? Gee, it probably doesn't even matter. Dollars to doughnuts this place is already swarming with Sequel Police. I can't believe I haven't seen any around. Oh well, so far so good. I should be thankful for small favors, and I guess I should start looking around. This can't possibly be more complex than Vohaul's asteroid!

First I need to figure out where I am—which window. Hmmm, it looks like there may be something there to the east. I'll bet I'm in the center room. Oh well, there's only one way to find out. I'll go east and see what happens.

East Window (Super Computer Entrance)

I was right! I did land in the center room. There's only one door leading into this area—except for the door on that thing over there. Hmmm, that looks like a tunnel entrance. I should know—I've certainly had enough experience with tunnels of all shapes and sizes.

Uh oh. Look what's standing guard over there in the corner! It's one of those darn Sequel Police! I knew there'd be a catch to my getting away so easily. I better just back myself out the way I came in and hope that guy doesn't hear a peep!

At least I know where I am now. I knew I was smart enough to figure this out. No one can fool Roger Wilco for long. That was the east room, and I landed in the middle room, just as I thought. And now to get on with the important business at hand—I'm on my way to the west room right now!

West Window (Time Pod Dispatch)

Aha, I was right! There is a west room! . . . Huh? Uh oh! I better be quiet—and very, very still. I knew there'd be more darn Sequel Police around here. . . . Hey, what the heck? . . . jiminy crickets, I've never seen anything like *that* before! How did he do that? Where did that spaceship come from? It just appeared out of thin air—the same way I did when I came through that . . . *time rip!*

That's it! I'll bet it's a time pod! That's how they travel through time! I have to get to that pod! I have to travel back to Xenon and warn everyone about the impending disaster! If I can sneak over there without those guys seeing me. . . . I have no other choice. I must take that chance. It's now or never. . . .

Time Pod Interior

Whew! I made it! Those jerks won't even know what happened. . . . On second thought, yes they will. They'll know exactly what happened, and they'll be after my hide in no time flat! What was that he said—he just completed a scan of the Labion sector of Space Quest II? Now what the heck is that supposed to mean? Gee, you would think this was some kind of adventure game the way those guys were talking.

If at first you don't succeed, try, try again.

Now all I have to do is figure out how to work this contraption. Hmmm, it shouldn't be too hard—not if those bozos can do it. The instructions say "Punch in correct time coordinates." Oh darn it, not a bunch of crazy symbols! Who do I look like—Indiana Jones?

How the heck should I know what the coordinates are? I need to get back to the Xenon of *my* time, but how? Oh well, I guess I'll just punch in something random and see what happens. Whoops! On second thought, before I do that, maybe I should write down the coordinates that are already there. Otherwise, how will I ever make it back to *this* Xenon?

OK, here goes nothing. . . . Hey, something's happening! I hear the engines firing, the turbines spinning, and the rocket boosters . . . winding down. Darn it, that didn't work. Geez, I could be here for 3000 years punching in coordinates! Something just has to work. Maybe if I take a more logical approach to this—I wouldn't have remembered the coordinates I just punched in anyway. Let's see, I'll enter another six symbols in a logical order and then see what happens. . . . OK, the engines are firing again; time and space are beginning to bend under the fibrilations of the time rip transfluxers. The turbines are spinning, and the rocket boosters are . . . going! Yeah! I did it! I must be a genius! Xenon of old, Roger Wilco is back!

Game Points Earned

Action	Points
Getting into the time pod safely	10

Space Quest X: The Planet Estros

Mission: Find the partial time pod coordinates. Save the La-tex Babes (and yourself) from the giant sea slug.
Total Points: 20

The Planet Estros

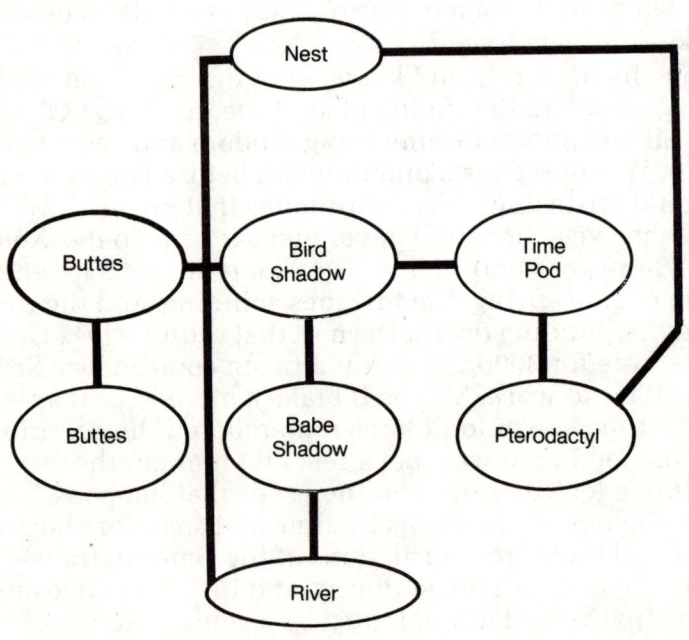

Inventory

chewing-gum wrapper

The Estros Buttes

Saaay, wait a minute—where the heck am I? This doesn't look like Xenon, at least not in this century—maybe prehistoric Xenon. No, no way. Just look at this crummy landscaping. I've never seen so many big buttes in my life! And look at all the narrow, winding little walkways in between. How are you supposed to find your way around here without falling off a ledge? This place is almost worse than Ortega, but at least it isn't falling apart.

Darn it, I wanted to go back to Xenon. Stupid time pod. That's probably some more of Vohaul's handiwork. Oh well, since I'm here, I might as well look around. Maybe I'll find a couple of loose rocks—something useful.

West of the Landing

Gee, this really is a narrow little pathway. I have to admit, I wouldn't want to take up permanent residence on this planet. Look at all that sedimentary rock winding its way down below. And those rock spires—they reach to the sky all around me. How poetic this place makes me feel in spite of the fact that it's ugly. Say, I wonder what the deal is with these wide, flat steps.

Ahhhh! What was that! M-maybe this really is prehistoric Xenon! That was definitely a prehistoric *something* that just flew over my head, and boy was it big! Just stay away from me, whatever you are! Maybe I should move away from this open area and explore a little more to the west.

Winding Steps

Hmmm, these winding steps must go all the way down to that river—stream—whatever it is. Oh brother, now look at that ledge. Boy, somebody must really think I'm stupid. No way I'm going to walk over there and risk falling off that ledge. A bunch of butte-heads built this place.

I am curious about that flowing river of aqua below. It's the only thing around here that isn't an eyesore. I think I'll just make my way very carefully down this winding stairway to see if I can reach the water.

Bottom of the Steps

Oh great, there's no place to go from here except up—or down if you're an idiot. Well I'm not stupid enough to fall off this ledge either. This place is the pits. I'm outta here. I'm going back to those wide steps I passed—maybe I can get to the bottom from there.

Wide Steps

Well, at least that giant bird, or whatever it was, hasn't been back. That really gave me the creeps. It doesn't look like there's much below here either, but I might as well climb down these steps and find out.

Huh?! What was that?! I saw something move behind that butte! It couldn't have been that bird again. No way—it was far too big. I'm beginning to dislike the feel of things around here. I think I'll just hightail it back to the time pod and try another set of coordinates. This is just another dead-end rock stairway anyway.

Who knows what thoughtless, chauvanistic evil lurks in the heart of one Roger Wilco? The Latex Babe shadow knows.

Well, there's my time pod again. It just sits there, waiting to take me to other time periods. Well, so far, Bucko, you haven't done a real swift job. Who do you think you are, anyway—the guardian of time? And I guess you think this is the city on the edge of forever. More like buttes on the edge of forever. Well I've got news for you, pal—that guardian of time theme is someone else's story, not yours.

Saaay, there's one other area I haven't yet explored. I could take a quick look down that other rock stairway to see if it leads to the water. What could possibly go wrong in that amount of time? I don't see any prehistoric birds flying around, so I'm sure a quick look will be safe; then I'll leave.

South of the Landing Site

Hey, what the . . . heck is . . . going . . . on?! Ahhhhhhh! Help! Heeeeeeelp! Let me go! Ahhhhhh! I don't believe this! It wasn't my imagination after all! I've been captured by a pterodactyl, a prehistoric flying reptile of the order Pterosauria from the Jurassic and Cretaceous periods! I thought they were extinct, but since I've noticed the characteristic wings consisting of a flap of skin supported by the very long fourth digit on each front leg, I know I'm dealing with a real, live pterodactyl!

Darn it! Why did I have to take one more look?! I was almost off this planet, and now I'm going to be dinner for this overgrown turkey! Ouch! Its talons are digging into my skin! Ahhh!

Oh no, I see a patrol ship below! It's the Sequel Police! They'll find my time pod! They'll know I'm here! . . . But what difference will it make anyway? I'm going to be lining the insides of this beast, and then eventually, after a day or two, I'll come out the other end as . . . something really gross!

In the Pterodactyl's Nest

Ahhhh! Ouch! Oh . . . my head. Oh I don't believe this is happening. At least I wasn't tops on the menu—yet. At least it wasn't hungry *now*! How am I ever going to get out of this mess.

Huh! The pterodactyl caught one of the Sequel Police! He's going to drop him here in the nest with me! Oh no, I hope he isn't armed! . . . Oooo yuck! I guess that's the conclusion to that sequel. Huh? What's he saying? . . . "Sorry, pal, there's not a darn thing I can do for you. All I can say is, glad it was you and not me." Gee, he's dead. He was after my hide like there's no tomorrow. Shoot, the way they're tampering with time travel these days, I wouldn't be surprised if there really were no tomorrow. How would it know which day to be, anyway?

Yes, that's you, Roger. Just another dropping in the bird nest of life.

OK, Wilco, enough is enough. I should be philosophizing about a way to get out of this nest. I wonder if that guy was armed. I don't see any weapons, but still, he may be carrying something of value on his person. Yuck. I just don't want to get any goo on my hands. OK, pal, here goes. . . . What? I don't believe it—a wad of used chewing gum stuck to a piece of paper. That clown was totally worthless.

Say, what's this though? There's some kind of code on the inside of the wrapper, but I can only make out three digits. Oh no, it's those darn symbols again. Don't these people know their ABCs? Hey, maybe these symbols are the coordinates back to the Xenon of my time! Whatever they are, I can definitely punch them into the time pod's keypad—assuming I'm ever in that time pod again, that is. I'll hang on to this wrapper just in case, though. For now, I need to find a way to escape before it's too late.

I see an opening in the corner of the nest—looks like that's my only way out. I'm not about to wait around here to be eaten alive by that cuckoo bird. Long way down or not,

I'm going to jump. Maybe I'll be lucky and land in the river.
Say your prayers, Wilco. . . . Here goes! . . .

At the River's Edge

Oh boy, I can't believe my good luck! I actually fell into the
river unharmed! Now to make my way back to that time pod
without being seen by the other Sequel. . . . Huh?! What
the. . . . What's going on? . . . "Hey! Who the heck are you?—
Boy, am I glad to see you gals! I knew this planet had its share
of beautiful buttes, but I never. . . .

In your dreams,
Roger! Hell hath no
fury like an Estros
Babe scorned.

"Huh? What do you mean shut up? . . . Hey, hold your
harpoons, I'm just an innocent traveler. . . . Huh? How do
you know my name? . . . Wait a minute, I don't recall ever
meeting anyone named Zondra. Believe me, I would remember *you!* . . . Oh no, you must have me mixed up with someone else. I could never leave you at the altar—any of you!

"Look, let me explain. You see, I'm not really from this
time period. You see, one time I was. . . . OK! OK! I'm going.
I'm going. Don't push. And quick sticking that harpoon in my
face—I've been through enough today as it is."

In the Underwater Cavern

Gee, for such nice-looking gals, they sure are mean. And she
called me fly-boy . . . *(sniff)* . . . which makes me think of my

little droid pal who is no more. And whenever I think of Fly-Boy, I think of Gladys . . . (sniff) . . . who's probably out wandering the streets of Xenon, screaming her head off every time she sees something move. These Babes are for the birds. I'll just stick to the Gladyses of the world from now on.

Gosh, I better get control of myself. I don't want them to think I'm a wimp—not with them being as tough as they are. "Uh, excuse me. Where are. . . . Oops, sorry. I was just asking." Where in the world is this place? I heard her say something about the planet Estros—that must be where we are. Estros, huh? Boy, I can believe it.

Latex Babes' Cavern

Well, it looks like we're here. Gee, they don't say too much. I guess Zondra's really steamed at me—definitely a woman scorned. If only they would give me the chance to explain instead of jumping to so many conclusions. I know one thing for sure, if I ever make it back to my own time period, I'll never get involved with any Latex-attired Estros natives.

I wonder what they're going to do with me. Uh oh, it looks like the one called Thoreen's going to be in charge of this. I wouldn't mess with her under *any* circumstances!

In the Torture Chair

OK, they have me seated in this chair. It looks like a barber's chair. What's she going to do, shave my legs? Heh heh . . . (gulp) . . . Huh? What happened to my boots—and my pants?! I'm wearing *cutoffs* now? Oh gee, please don't let anyone ever find out about this—EVER! I'd be the laughing stock of Xenon—of the whole universe. . . . Oh no, now what is she doing? That looks like a . . . (gulp) . . . razor! It is! It's an EpiRip ladies' shaver! She IS going to shave my legs!

"Ack! Oh no! You wouldn't do that to me, would you?" Those darn things, they rip hair right out of its socket. And they hurt really bad—not that I've ever used one, but I've heard all about them. . . .

The Sea Slug

Huh?! What did she say?! "Hey, did you just say *sea slug?* . . . Y-y-y-yikes!" That's exactly what she said, and I *hate* slugs!

Those nasty, slimy things. I wish I had a gallon of salt to pour on it. Oh my gosh, it's heading my way!. . . . I can't get out of these restraints! "Hey! You're not going to just leave me here, are you?! Wait a minute! I don't deserve this!"

Those wimps. They sure talk a tough game, but when it comes to something real, they all run and hide. . . . Oh no! Its tentacles are wrapping around my legs. . . . I've got to reach that button with my thumb. . . . Ah! I did it!

Poor slugger, all he needed was a little air.

Think fast, Wilco! Think fast! There's no time to waste! . . . Those oxygen tanks! I saw this in a picture once—the guy was about to be eaten by a giant shark when he threw an oxygen tank into its mouth! . . . Ahhh! Here goes. . . . It's my only hope!. . . . Ahhhhh!

Ohhhh! Whew! I did it! It's gone! I destroyed the sea slug! "Hey girls, it's OK now—you can come out from hiding. I've saved the day! . . . Well, Thoreen, I guess this means you'll be taking back that thing you said about my not being a real man. . . . Aw, gee. It was nothing! . . . Anything for you and your friends.

"Say, Zondra, now that we've made up, let's talk about this situation between us. . . . What? Well, uh, wait a minute now. I was really hoping for the opportunity to apologize properly. Couldn't we go to some nice, quiet, private place—say your quarters—and talk about this? . . . What? SHOPPING? You mean there's a mall close by? Hey, all right! Who wants to talk when there's shopping to do?! What are we waiting for? SHOPPING MARATHON! Let's go! . . ."

Game Points Earned

Action	Points
Getting the chewing-gum wrapper	5
Getting the oxygen tank	5
Killing the slug	5
Zapping the sea slug's tentacles	5

Object Locations

Object	Location
Chewing-gum wrapper	Dead Sequel Police

Space Quest X: The Galaxy Galleria Mall

Mission: Earn enough money to buy what you need. Find your way back to *Space Quest I.*
Total Points: 53

The Galaxy Galleria Mall

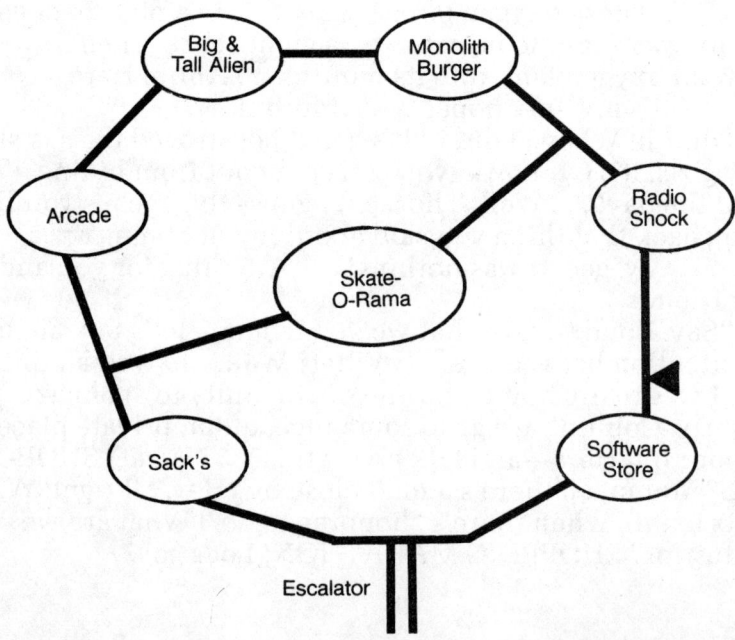

Inventory

autoteller card
buckazoids
cigar
dress and wig
hint book
pants and boots

Mall Entrance

"All right! The Galaxy Galleria mall! I didn't even know this place existed! Boy, this is swell. You girls really know how to have fun! And look at that great Skate-O-Rama! . . . Hey, wait a minute! Where's everybody going? . . . Sale? Where? . . . Hey! . . ." Gosh, I guess they all forgot about me when they spotted those great sales. Oh well, I still have a couple of buckazoids in my pocket—I might as well spend them here!

Oh gosh, Zondra dropped her autoteller card, and now she's nowhere in sight. I'll just hang on to it for now and give it back to her later—if I see her, that is. Hmmm, that looks like an autoteller machine over there next to the software store.

Saaay, I wonder if Zondra would mind if I borrowed a couple of buckazoids from her. I mean, I'll pay her back later. It's not like I'd be stealing it or anything. Besides, I just saved her skin from that sea slug, and I think that's worth a little something—even if she does think I left her at the altar.

Boy, that software store is packed. What in the world is all the commotion about? Someone really famous must be in there! I can't believe this crowd. One thing's for sure—it doesn't look like I'm going to get inside anytime soon. I'll just ask someone what's going on.

"Excuse me, what's all the excitement about? . . . WHAT?! You mean to tell me those two ingrates are in there signing software? I rescued them from Pestulon and how do they repay me? By jumping ship at the first lousy job offer that comes along. Well, no thank you. I wouldn't buy their sorry computer games if they were the last software designers in the universe.

"What do you mean, shut up? You shut up! All I did was ask a simple question. . . . And what are you looking at, anyway? Haven't you ever seen cutoffs before? They're all the rage on Estros. . . . Yeah, well you couldn't exactly wear them with those legs. . . ." Stupid jerk. He probably thinks the Two Guys are the greatest things since Ronko came out with that stud and rhinestone machine. I can't believe those Guys are here. And how the heck did they get to this time period? Oh, I get it. They've probably spent their entire lives slaving for that Sierra company—the same company who refused to give me, an experienced janitor, a simple little cleaning job. Yeah, that was really some fine how-do-you-do. I could have avoided all of *this* if they had just given me a job. Oh well, I guess that's all sewer in the tank now.

Still, it depresses me to think about it. I could really use some cheering up about now—after the way this day has gone. In fact, I believe I'll just put this little autoteller card to use. That ought to help cheer me up. All I have to do is pop the card in the slot, and. . . . What?! Identification negative? How does that blasted machine know I'm not Zondra? Maybe she got a haircut or something. Darn it, that figures. Nothing is working out right today. I look like a dang hobo in these cutoffs.

Maybe I can find a good sale on men's clothing somewhere. . . . Oh boy, is that what I think it is? It is! Forget about clothes right now! Monolith Burger, here I come! I can't wait to sink my teeth into one of those juicy burgers! I'm starved!

At the Monolith Burger

"Hey, what's the big idea? You can't throw me out! I see the sign you old snoutface." Great, I can't even get into a lousy fast-food joint without the right clothes. Since when has the Monolith Burger become so strict about patron attire? That's the stupidest rule I've ever heard. . . . Darn it. I might as well try to find a men's clothing store.

Big and Tall Alien Store

Hmmm, what's this? "Hey, Beanpole, watch where you're going—you just knocked out half the store sign! . . . Oh yeah, you wanna make me?" Jerk. I can still make out the name on the sign, though. It says, "Big and Tall Alien." Well, I guess that means me. I mean, I'm no alien but I am kind of big and tall—but not *too* big or *too* tall. Us swashbucklers have to be just the right size.

Yes, this is definitely my kind of store, so maybe I'll have some good luck in here. I'm beginning to feel a little embarrassed, walking around in these cutoffs with no shoes. Someone might think I'm a hillbilly or something—like that guy coming out of the store. Geez, I wish he'd quit staring!

Captain Kirk may have originated the post-scuffle, shredded-shirt, torn-pants look but, Roger, this is SPACE QUEST. Get some clothes.

"Hey, Fungus-head, who are you calling Jethro? The name's Wilco—Roger Wilco—and don't you forget it!" Ooops! I better be quiet. Some of Vohaul's cronies might be hanging around within earshot. I better keep my mouth shut and just get on with the business of buying clothes.

"Excuse me, sir. . . . Yes, I need to buy a pair of pants, and I'd like some boots too. Do you have anything that's . . . er . . . on sale? Oh, OK. Hey, those look great! Thanks, I'll try them on right now. Which way to the dressing room?"

Wow, do these togs look great or what? No wonder Zondra fell for me so hard. You can see every sinuous inch of brawn in my legs, and these boots are oh-so cool. Boy, I could give a lot of heroes a run for their money. I'm sure the sales clerk will agree.

"Say, pal, what do you think? Are these me, or what? . . . And they're how much? Great! Sounds like a good deal to me. I'll take the pants and the boots. . . . No, thank *you*, and have a nice day."

On the Moving Sidewalk

Gee, that was a great sale. I feel like a new person again, but I still need some more buckazoids if I'm going to do any more shopping. I'd like to go back to that software store if those two jerks would hurry up and leave. In the meantime, I guess I have just enough buckazoids to buy a nice big Monolith meal. And we'll just see if that snoutface tries to kick me out this time.

Back at the Monolith Burger

"Excuse me, sir. Is my attire acceptable to you now? Oh, well good then. I'd like to order some food. . . . Well you're an employee aren't you? Can't you get it for me? What's your job, anyway, picking your. . . . Oh, all right, all right! You don't have to yell. . . . A job? Well, yeah. As a matter of fact, I could use the extra buckazoids right now. I still have some shopping left to do. . . . Well OK. I'll get started right away."

Oh brother, this is going to be a cinch. What could possibly be difficult about fixing a bunch of Monolith burgers—except to refrain from eating them.

Gee, Roger. You can't always expect a free ride. Sometimes you have to work for a living.

OK, here we go—here comes the first burger patty. Let's see, first I put the lettuce on, then the relish, then the mayo, then the mustard, then the ketchup, and last but not least—the bun. I did it! I knew this would be a breeze, but what a slow machine. Gosh, he's only paying me one buckazoid per burger. Can't this conveyor belt move any faster?

Here comes the next one—lettuce, relish, mayo, mustard, ketchup, and the bun. Shoot, I could be here all day with this boring job. I know—I'll quit when I get to 50 buckazoids. That should be enough for me to make a few more purchases. . . .

Finally! I'm at six measly buckazoids and the conveyor belt finally decides to speed up. It's about time. OK—lettuce, relish, mayo, mustard, ketchup, and the bun. Here comes the next one—lettuce, relish, mayo, mustard, ketchup, and the bun. . . . And the next one. . . .

Hey, wait a minute! The conveyor belt is moving a little too fast now. How am I supposed to keep up with all these patties? That jerk must have turned up the speed. OK—lettuce, relish, mayo . . . mayo . . . darn it, forget the stupid mayonnaise! No one will even notice. . . . Let's see, mustard, ketchup, and the bun! Whew! . . . What?! What do you mean it's a reject? Hey, everyone doesn't have to have mayonnaise on their Monolith burger!

Geez, I can't believe I'm standing here arguing with a machine. Here comes the next burger . . . mayo, ketchup, relish. . . . Oh, darn it! Another reject. . . . "Hey manager, this machine's messing up! I can't keep up with this crazy thing! It keeps rejecting my perfectly good burgers! So what are you going to do about it? . . .

Outside the Monolith Burger

Darn it, I don't believe it! That jerk fired me, and all I made was a lousy ten buckazoids—diddley squat! How am I supposed to buy anything now? . . . Hey, he threw his cigar at me!

Hmmm, I think I'll swipe that cigar. I may have a use for it later. Yeah, maybe the next time I walk into to that jerk's Monolith Burger joint, we'll see what pops up out of one of his burgers. Heh heh heh. That'll show him not to fool with Roger Wilco. Now come back here, you wad of tobacco. . . .

I got it! I got the cigar. Eeew, it's yucky and nasty—especially considering where it's been—yuck, in that ogre's mouth. Blah! I could gag just thinking about it!

Well, now where can I go? I see an arcade room a few doors down, but who wants to fool with a bunch of boring arcade games. So what if I was once the best *Astro Chicken*

player around? I can't get interested in those things anymore—especially now that I know the Two Guys are making
a fortune off the suckers who play them. Boy, ripoff city.

I passed a RadioShock store earlier. Maybe I could find a
cord or something in there to go with my PocketPal computer. It isn't doing me a heck of a lot of good in my pocket.
At least I have that Energizer battery for power, though.

RadioShock

"Hi there. . . . Yes, I'd like to take a look through your catalog." Wow, this is pretty neat. How convenient to be able to
order straight from this automated catalog. I wonder if they
have anything I can use.

Hmmm, *ReShrinkwrap 2000. . . . Is that game new or
used? Only you'll know for sure.* Dealers only? Oh I get it. So
that's how some of those seedy software joints stay in business. They reshrinkwrap used software. I'll bet half of it
doesn't even work! What a bunch of swindlers.

*Buying just
wouldn't be pru-
dent at this junc-
ture—not that you
can afford it now
anyway.*

Hmmm, now here's something interesting. An adapter
for my PocketPal. Wow, that's exactly what I need! But wait
a minute—I don't even have anything to adapt it to. Saaay,
this stuff is pretty darn expensive anyway—1999 buckazoids
for one lousy adapter? Hah! I'd have to be desperate to buy
that!

In fact, everything in this store is expensive—and half of
it's out of stock or discontinued. What kind of setup is this,

anyway? I thought automation was supposed to make everything better. Boy, what a joke. This company can't even keep its bunch of gadgets in stock, and they don't even have anything good. Some dandy corporation this turned out to be. I'm bagging this joint—el pronto.

Outside RadioShock

I've got to find a way to get by that autoteller scanner—but how? If only I could disguise myself. . . . That's it! I'll buy a couple of ladies' things—like a dress and some heels—and then I'll use that autoteller card. Yeah, that's it! That's what I'll do. And I know just where to go for the goods—Sack's. That's the best women's store in the universe!

At Sack's

Wow, look at those mannequins! Boy are they hot! If I didn't know better, I'd ask one out for a date! Uh oh, here comes the sales robot. . . . "Uh, yeah. I'm looking for something to wear. . . . I mean something to *buy* for my lady friend to wear. . . . Well, she's kind of a big girl. I'd say I'm close to her size. . . . Oh, that looks perfect! A black mini—those little black dresses are really the thing now, aren't they? . . . Oh, hee hee, I'd be much too embarrassed to try it on in place of her. Oh, well—if you insist—I guess I could. . . . Oh perfect! I'll . . . she'll love the wig, too. Thanks, ma'am. I'll be out in just a minute. Now don't peek!"

Hey, I make a pretty good-looking woman, if I say so myself. Just don't get too used to this, Wilco. Truth be known, I'm a sorry sight for sore eyes. If *anyone* on Xenon ever found out about this—I'd be ruined forever. Especially Gladys— she'd *never* go out with me. But hey, what am I worried about? This is the future. I'd have to go back to the present in this dress for anyone to see me. Well, I guess I should go out and face the sales robot now. I just hope this outfit isn't too expensive.

"Uh, excuse me. I thought I would . . . uh . . . wear this out in the mall—after I buy it, of course—because . . . uh . . . I want to . . . uh . . . go play a joke on somebody! Yeah, that's it—I'm going to play a joke on one of my friends. He's had it coming for a long time. Say, how much is this going to cost me anyway—for the whole outfit, I mean? HOW MUCH?

Gee, well it's not that I can't afford it or anything. Maybe I'll just think about it for awhile as I browse. Just don't sell it to anyone else, OK?"

Gee whiz, how embarrassing can you get? I'll bet my face turned a thousand shades of red, even if she is just a robot. And what was that "knowing" little wink supposed to mean? Oh, what the heck. What difference does it make now? I can't believe I don't have enough buckazoids for that dress and wig. Now I'll have to crawl back over to the Monolith Burger and beg that guy for another job. This isn't going to be fun.

Back at the Monolith Burger

"Yeah, it's me again. Look, that conveyor belt messed up last time. . . . Well if that's what they all say, maybe you should listen up and get the thing fixed! How do you expect to keep employees if that conveyor belt makes it impossible to do the job? . . . Look, all I'm asking for is one more chance. I'm sure I'll get the hang of it this time. . . . You will? Great! Thanks, sir. You won't regret it."

Snoutface. I'd like to stick that reject button up your cavernous nostrils. OK, here we go again. I *hate* this job, and I know the same thing's going to happen all over again. Maybe this time I'll figure out a way to outsmart this stupid machine. . . .

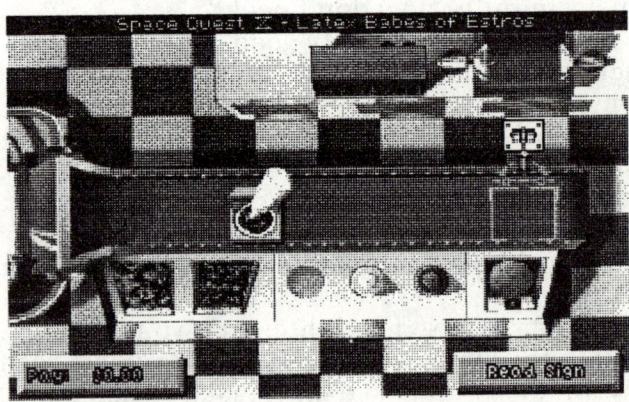

The best way to approach this job is with your hands. Don't mouse your way out of it.

... Lettuce, relish, mayo, mustard, ketchup, bun. ...
Lettuce relish mayo mustard ketchup bun. ... Lettucerelish-
mayomustardketchupbun. ... Let-rel-may-mus-ket-bun. ...

All right! I did it! I have enough buckazoids now, including the ten I earned in here earlier. Now I should have
enough left over after buying the dress and wig to get some
really neat software!

"Hey Snoutface, pay up. You owe me a bundle of buckazoids. ... Look, Mr. Ed, I'm personally going to stuff your
face in every condiment on that counter if you don't chuck
up the dough right now—got it? I don't give a flying fig about
the rest of those burger patties—they can reject all the way
to the next galaxy for all I care! Now give me my buckazoids! ... Thanks. And have a nice day."

Back at Sack's

Whew! That was a pain, but at least I got the bucks to buy
these clothes. Now I can get to that autoteller and really rack
up. Gee, Zondra and the rest of the Babes are bound to have
left by now. Oh well, I'll pay her back one of these years.

"Hi ... uh ... yes, I've thought about it for quite a while
now. I do believe I'll take that black dress and the wig. My
friend is going to heehaw when he figures out who I am, ha
ha ... OK, I'll just do a quick change in the dressing room
... uh ... I appreciate your holding these things for me.

*As they say, clothes
make the man.
And that cute little
swing in your stride
is a real addition,
Roger. Let's hope
the ATM has an ad-
dition or two for
you.*

"OK, now that I'm dressed, I'll be glad to pay you. . . .
There you go. Oh yes, I'll be back for sure. . . . No, thank *you!*
It's always a pleasure shopping at Sack's . . . uh . . . I mean
my girlfriend, Zondra . . . er . . . Gladys . . . this is her favorite
store. Thanks again."

There she goes with that "knowing" wink again—like
we have some little secret together. Hmmp. Who cares what
she thinks. I'm more of a lady than she'll ever be! Hmmp.

At the Autoteller

Boy, I thought I'd never make it to this teller in these heels
and this dress. . . . Hee hee hee. . . . The darn hairs from this
wig are tickling me like crazy, and . . . hee hee hee . . . I can't
stop giggling. With the way I have to walk in this getup and
. . . hee hee hee . . . the way I keep snickering, no one will
ever suspect there's a real man under this wig.

OK, autoteller, let's get it right this time. . . . Identity positive! All right! Now, how much do I need? Hah, that's easy—
clean out the house, pardner. Now we're in business! I've got
a pocket full of buckazoids, some great new pants and boots,
and now I can buy myself a neat computer adventure game
at the software store.

Now if I can just get back to Sack's without busting my
as . . . er . . . without breaking my back. *And,* if I can get back
before one of these slimeball aliens—with their fat, fleshy
tongues hanging out of their mouths—makes a pass at me. I
mean, I know I make a nice-looking woman—heck, I'd make
a nice-looking anything—but these creeps don't have to act
like they've never seen a skirt and two legs before.

Final Sack's Stop

Whew! I made it. "Hi there. Just me again. . . . Yes, my friend
was very amused . . . ha ha . . . . Yeah, he was real amused.
He almost asked me out! Can you believe it? . . . ha ha . . . .
But boy was he embarrassed when he found out it was me!
Yeah, that was the greatest, but I'm ready to get back into my
own clothes now. I'll just, uh, give this outfit to my sweet-
heart some other time. Well, thanks again. You have a very
sacks-y store here, ha ha. Bye now."

Boy am I glad that's over. It feels great to be back in a
man's clothes. Whew, I hope I never have to do that again.

That was enough to give a guy a real identity crises for crying out loud. Now, where should I go—to the software store, of course.

At the Software Store

Wow, this place really cleared out, and I'm glad those Two Jerks are gone. They ought to be called the Two Nerds from Andromeda.

"Hi there. I thought I'd just browse a bit. . . . No more Sierra software? Well whadaya know—this must be my lucky day. Say, what's that big box over there? . . . The bargain bin? . . . Well, let's see. Maybe I'll find a good deal in there."

"Saaay, what kind of software is this? I've never heard of this stuff." *Where in the Heck Is Hymie Lipshitz? SimSim— The Simulator Simulator, Dacron Danny*—boy, these guys are ripoff kings. *Checkers Construction Set*—well that one sounds pretty interesting. *King's Quest XXXXVIII*—oh gee, give it up, Sierra. Give that poor old king a rest.

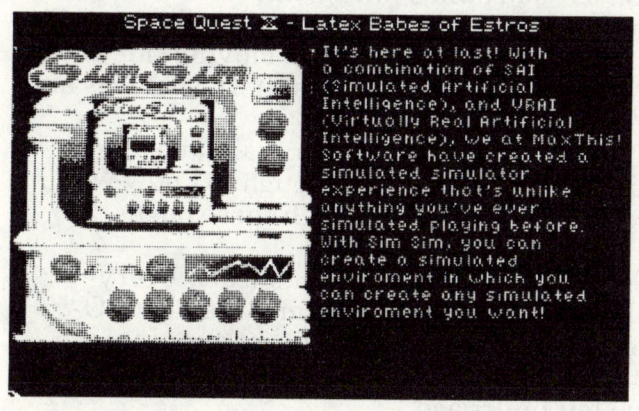

You know a good sale when you see one. But remember: it isn't a bargain unless you can use it.

Hey, what's that on the bottom? It looks like a hint book. . . . It is a hint book: *Space Quest IV Hint Book.* Hmmm, this sounds like something that could be useful to me—even if it is a Sierra book. I promised myself I would never buy anything from that Earth company after they snubbed me over that darn janitor's job. . . . Well, why not let bygones be bygones? After all, I am in a space quest of sorts. I mean, I'm in space and I'm on a quest, and this book might really turn out to be helpful.

Gosh, that reminds me—I still have to find Sludge Vohaul and stop him! I was having such a good time at the mall that I completely forgot about Xenon! I can't believe I did that! Well, that's it—Sierra or no Sierra, I'm buying this hint book.

"Excuse me, sir, I'll take this *Space Quest IV Hint Book*. . . . Here you are. Yeah, thanks." I sure hope this book has some good advice for space travel. Maybe it will even help me figure out how to get rid of Vohaul! Nah, that's asking too much.

Outside the Software Store

Gee, I'd like to look through this hint book, but I'm afraid the marker will run out even if it is battery operated. Those cheap things Sierra used to sell with their hint books never lasted. They put just enough ink in them to let you read almost all the hints, and then WAM!—no more magic ink.

Well I don't trust this battery-operated ElectroEraser either. Just as soon as I start uncovering hints line for line, like I used to do with the markers, the battery will run out. I know it just as sure as I'm standing on this moving sidewalk. So what if I cheated and read all the hints first? That was my business!

No sir, I'll just wait until a really important question pops into my head. At least I'm smart enough to know which hints to uncover. They can't trick me with those bogus questions.

Now where should I go? I've been practically everywhere in the mall except for the Arcade and the Skate-O-Rama. I don't even know how I'm going to get out of this mall. I sure can't go back the way I came in—Zondra and her friends must have left hours ago.

All these buckazoids and nowhere to spend them. I'm not even in the mood to shop anymore now that I've started thinking about Xenon again. But what good does it do to think about Xenon? I can't do a darn thing about any of it right now. I'm really getting depressed. . . . This isn't good. I need a pick-me-up for sure—something to really light a fire under my feet.

Well, maybe I could play an arcade game. I used to really get into *Astro Chicken*—not that I'd fool with that ScumSoft game again. Maybe there are other games that would interest me—some that would get my adrenalin going again. I guess it's worth a try.

At the Arcade

Yep, looks pretty much like a typical arcade to me. I've been avoiding this place like the plague all day, and now I'm finally here. Hey, I don't believe it! All my old favorites—excepting *Astro Chicken*, of course—are still in commission. Wow, I have an entire smorgasbord to choose from: *Ghetto Blaster*, *Dweeb Hunter*, *Choke-'n-Croak*, and a whole lot more!

Saaay, wait a minute. What's that? I don't believe it! *Ms. Astro Chicken?* Aw, you've got to be kidding, Guys. That's the stupidest thing I've ever seen—next to *Astro Chicken*. I'm not about to play that cheating game. No way. Not me. I've had enough of those Two Guys' arcade games. . . . Well, maybe I'll try it just once. . . .

Hey, this is really great! I can already feel the old energy coming back. Yes sir, it's time to put the Arcade Master back in the ring. . . . Huh?! What's going on over there? . . . *(gasp)* . . . A time pod! Oh no, it's the Sequel Police! Somehow they've managed to catch up with me! Great balls of fire, what am I going to do?!

Astro Chicken was nothing compared to this chick.

One of them went out into the mall, but the other one . . . he's coming around this way. Maybe I can sneak out without him seeing me. . . . Y-y-y-yikes! He sees me! He sees me! Now what do I do?! I nearly got blasted to eternity! The

Skate-O-Rama! Maybe they'll confuse me with all the skaters! It's worth a try!

In the Skate-O-Rama

Whoa-whoa-whoa-whoooooooaaa! I had forgotten how it felt to be in an antigravity zone. . . . Whoo! Yippee! That was fun! . . . I'm flying! I'm free as a bird! I'm . . . I'm being pelted with laser blasts from the Sequel Police! Yikes! I had hoped those laser beams wouldn't be equipped with anti-antigravity devices. . . . Looks like that was wishful thinking on my part. . . . Whoo! That was a close call.

"Hey, watch where you're going next time, young man!" Darn teenagers! They could care less who they sideswipe. . . . Oh no, those guys are really closing on me. They're right on my tail. If I could just get a little momentum going and get up in the air a little higher. . . . Whoa, that was a close one! How does Vohaul recruit these clowns? He must pay them a fortune to do all his dirty work—either that or they're brainwashed. Yeah, that's it. They must be brainwashed. . . . Whoa! Another close one! I'm almost at the stairs. . . . Just a few more inches . . . . And now to go back down as quickly as possible. If I can make it to that gravity zone while they're still high up in the air. . . .

I did it! I'm out! Where's the arcade? I've got to get to that time pod before the Sequel Police get out of the Skate-O-Rama! It's my only chance—otherwise, I know I'll never leave this mall alive, and Xenon will be lost to Vohaul forever. He'll have won. . . . There it is! There's the time pod, and here I come!

In the Time Pod

Whoa! This is getting too close for comfort again. I've got to get the heck out of here now. . . . Hey, those must be the mall coordinates. I better make a quick note of them while I can—not that I plan to come back here or anything, but gee whiz, you never can tell.

Darn it, now where should I go? I can't stay here, that's for sure. Think, Wilco, think! . . . Shoot, I haven't a clue. Wait a minute—that's it! That's what I need—a clue! A clue! A clue, a clue, a clue! The *Space Quest IV Hint Book*! It may be just an old bargain bin-buy, but at this particular moment, it may also be my only hope.

Hmmm, let's see what this says. . . . "Strawberry is to Sky as Fire Engine is to. . . ." Gee, how should I know? What does the answer say? Blueberry and India? Wait a minute; what kind of clue is that? That's no help! Darn it! That's not a real clue! It's a trick question! I can't believe I fell for that!

Boy, I'm really embarrassed. I'm glad no one saw that! Now, I have to find a legitimate clue. Hmmm, maybe this is one. If I'm at the mall and I don't know where to go. . . . Heyyy, wait a minute. What's that supposed to mean? Oh, ha ha—this is really some time to be making a joke. My life is hanging in the balance and these guys are giving wisecrack answers. . . . Here it is! It says I should punch in the following coordinates to—Ulence Flats?! Oh, right. Like I'm really going to Ulence Flats flats again. I *never* want to set foot in that crummy little town again. This is obviously a trick answer, but I might as well use the coordinates. I've got nothing else to go on.

Wait a minute! It only gives three. . . . How stupid can you get? Where am I supposed to get the other three? . . . Wait a minute! I still have that CHEWING-GUM WRAPPER! I knew it would come in handy sometime! OK, we're in business now and not a second too soon. I'll just punch in these symbols as fast as I can and, hopefully, I'll be on my way back to the Xenon of old. Where else *could* I be going?

Game Points Earned

Action	Points
Buying the dress and wig	5
Buying the pants and boots	5
Buying the right hint book	5
Getting a job	3
Getting the ATM card	2
Getting the cigar butt	5
Obtaining money from the ATM	10
Putting normal clothes back on	3
Refusing a job	−3
Safely escaping the mall	15

Object Locations

Object	Location
Autoteller card	On the floor at the mall entrance
Cigar	On the moving walkway
Dress and wig	Sack's
Hint book	Software store
More buckazoids	Autoteller; earn from Monolith Burger
Pants and boots	Big and Tall Alien store

Space Quest I: Ulence Flats, Kerona

Mission: Get a book of matches from the bar; then return to Space Quest XII.
Total Points: 20

Ulence Flats, Kerona

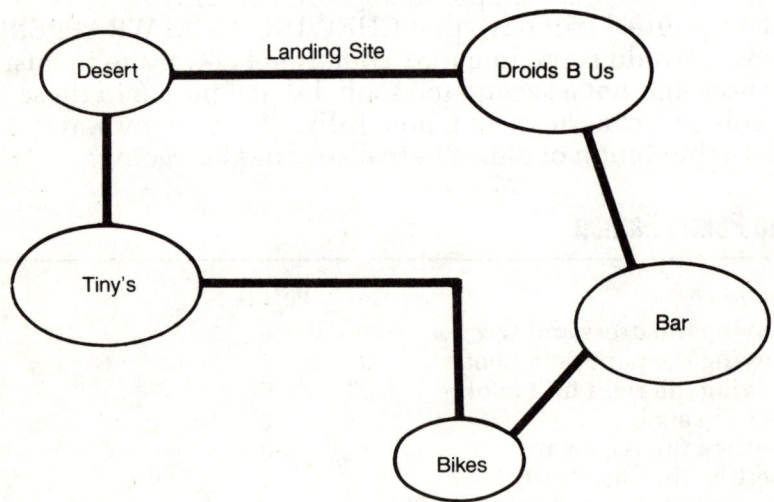

Inventory

matches

At the Landing Site

Say, this place looks awfully familiar—and I mean *awfully*—
but it sure isn't Xenon. It's so flat and two-dimensional.
When have I been here before? And what in the world is that
dome-shaped building? . . . *Droids B Us!* No, it can't be! *I* can't
be! *This* can't be—Ulence Flats?! Oh gee, talk about going
back in time. I never wanted to see this place again. I thought
that was just a joke in the hint book.

   And gosh, it . . . *(sniff sniff)* . . . brings back so many
memories. This is where I first found my little droid friend. I
remember it all so clearly. I told the salesman I would take
the little bargain droid, and from that moment on, his name
was Fly-Boy . . . *(sniff sniff)* . . . Poor old Fly-Boy. Now he's
probably just an evil death droid hunting down the few remnants of life left on Xenon. Well, I can't think about that now.
I'll drive myself crazy with grief. I need to concentrate on
something else.

*Come again? Isn't
this where you
swore you would
NEVER return?*

Say, I wonder if there's another five buckazoids on the ground behind the Droids B Us store! I remember how finding those buckazoids really turned my luck around when I was so down and out. I couldn't even buy a lousy beer. . . . Nope, no buckazoids on the ground this time. Oh well, it was worth a try. Speaking of beer and buckazoids, I wonder if they still brew their beer in a cauldron. I see the bar is still standing, but how the heck that guy stays in business is beyond me. Lousy beer, lousy service, and a lousy band—well, the band was OK, I guess.

I see Tiny's is still standing, too, though no one's around. No one's around anywhere. This place is pretty deserted. I do hear music from the bar—sounds like that same old band. Hmmm, and judging by those speeder bikes parked out front, I'd say there are definitely people in there. I think I'll go in and see who's there—probably the same old boring patrons.

In the Bar

Yeah, just like I thought—same old band, same old bartender, and probably the same old lousy service, although someone's keeping the bartender pretty busy right now. Who are those bland-looking guys? Holy hercules card, it's a pack of Monochrome Boys. What are they doing here in Ulence Flats? Those guys are bad news. . . . Uh oh, I have a feeling they're in the mood for a little action. I don't like the way they're looking at me—like I just stepped out of a color monitor.

Hmmm, are those matches over there? I never seem to have any when I need them. Maybe if I just casually walk over to the bar and pick up the book of matches, they'll allow me to leave peacefully.

"Uh, excuse me fellows. Just popped in to bum a match . . . . I . . . uh . . . enjoy a Keronian cigar every now and then, and. . . . Huh? . . . Eeew, did you actually eat that fly? Ugh. That's so gross. . . . Look, all I want is the book of matches. . . . Look, big guy, I never said I was hot stuff just because I'm in color and you're not. Heck, but now that you mention it, you could stand a pixel or two of red—for your neck, that is. Nyak-nyak-nyak. . . . And what are you bobbing your head up and down about, Numskull? . . . Y-y-yikes! Hey,

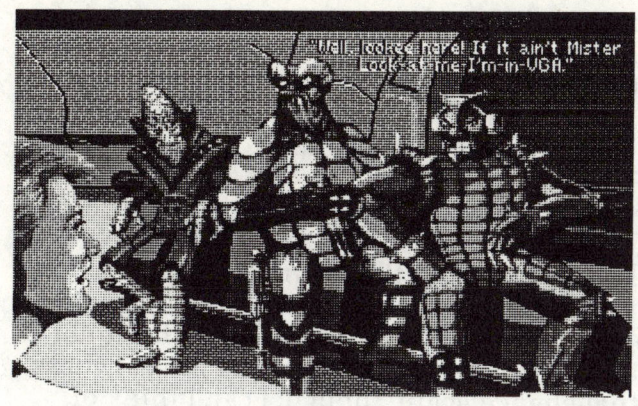

"Well, looee here! If it ain't Mister Lookee-here-I'm-in-VGA."

Hercules is a mis-nomer—they only look tough. You can outmaneuver a bunch of mono-chrome morons any day.

now wait a minute boys, I was just picking! . . . Hey, you started it! . . .

Outside the Bar

Ouch, that smarts! What a bunch of imbeciles. Darn it, I want those matches. They think they can kick sand in my face just because they're bigger than me and there are more of them. Well, I can kick something too—like their puny speeder bikes. I always have hated bikers.

Hah! That ought to teach them! Whoops! I better get out of here fast. Those guys will be on my tail in no time flat. . . . I need a place to hide! Tiny's! I'll hide behind Tiny's place! . . . They'll never find me there. . . . Whew! . . . *(puff puff)*. . . . The coast is clear. Now to get back over to that bar, swipe those matches, and get the heck out of Dodge . . . er, Ulence Flats!

At the Time Pod

Great! I managed to get the matches without incident. Those Mono Boys must have taken off over the desert looking for me. Tough luck to you, schmuckos—I'm out of here, just as soon as I can get in my time pod, and. . . . Huh? What the. . . . Oh no! The bikers! They're still after me! There's one headed my way right now. I'll have to make a dive for it! . . . . Whew! I made it! "Hey you boring black-and-white, antiquated piece of computer line art! Eat pixels!"

Boy I timed that right! Now to get the heck in that time pod before one of those jerks shows his face again. I just want to *leave* this place. Please, may I *never* return here again!

But where should I go? I've tried everything I can think of to get back to the Xenon of my time, but nothing works. Maybe I can't go back there—ever! Maybe I'm just a victim of time travel, destined to be lost in these space quests for ever! Oh, woe is me! What am I going to do? I don't know how to go home to my own time, and I don't know how to get through the laser tunnel leading into the future Xenon's Super Computer. Maybe I should go back to the future and try it again.

I never got the opportunity to check out that time pod dispatch area because it was always swarming with Sequel Police. Heck, I barely made it into the time pod alive. I guess I'll have to take that chance again, though. I can't think of anything else to do. I need another clue, but this cheap, bargain-bin hint book is no help at all. Should have known. I'll just pray that I find a way into the Super Computer and that the Sequel Police are off trailing someone else. Now, where are those coordinates I wrote down for Xenon? . . .

Game Points Earned

Action	Points
Evading the bike	5
Getting the matches	5
Kicking the bikes	5
Using the hint book code	5

Object Locations

Object	Location
Matches	In the bar

Return to Space Quest XII

Mission: Find a way into the Xenon Super Computer complex. Learn how to use your PocketPal.
Total Points: 35

Xenon Super Computer Complex

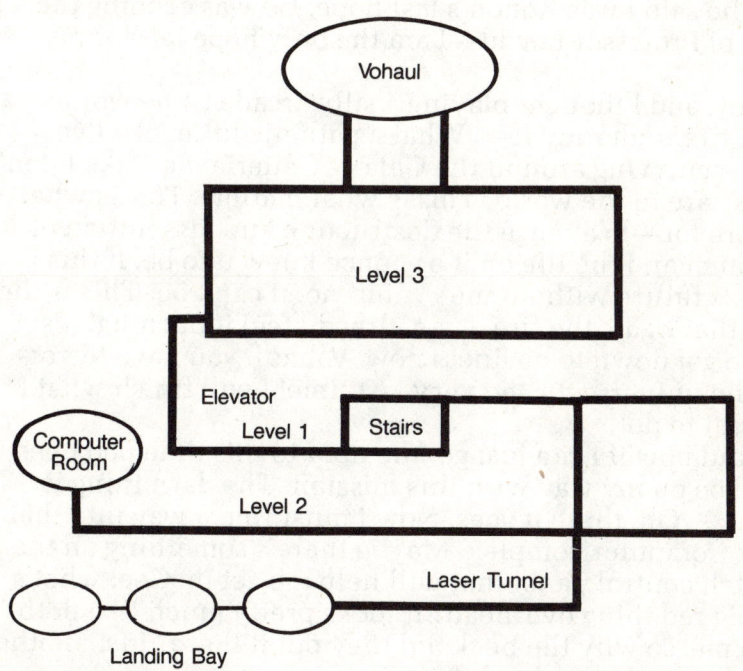

Time Pod Dispatch Bay

At least this time pod works. Heck, I could have ended up on Ortega for all this thing is worth. But I didn't—I made it back to Xenon, and what a depressing sight it is, looking out over those once-proud cities now so fragile and crumbling. Ruinous reminders of the darkest plague that's ever engulfed the universe.

The Bank of Xenon was once one of the strongest financial institutions in the galaxy, and now it's just a desolate old building, empty and forlorn. It stands there waiting, cold and empty, for someone to pass by and reopen its doors—for someone to bring life and warmth back into its idle halls. Waiting, waiting, waiting for someone to come by—other than that cyborg dragging his leg. No one will ever come, though, not for as long as those death droids patrol the ruins . . . (sigh). . . .

I must be brave. I must be strong. I can't cry out for help because there isn't a soul to hear, and besides, I don't want to draw unnecessary attention. That Time Ripper was right when he said I was Xenon's last hope. He was echoing the words of Professor Loydd—I am the only hope left for my planet.

Boy, and I thought playing a silly arcade game would light a fire under my feet. What a pathetic little joke I've been—scurrying around the Galaxy Galleria mall like I don't have a care in the world. *This* is what matters. *This* is what I'm here for—to avenge the destruction and dissolution of my planet and the life on it as I once knew it to be. If this is Xenon's future without me . . . but no, it can't be. This is the spark that lights the fire. I can already feel it burning! It's time to get down to business now. Vohaul, you have terrorized the universe for the very last time! I will finish what I was sent to do!

And now if I can just get the door to this time pod open . . . I'll be on my way with this mission. The darn thing is stuck. . . . Oh, there it goes. Now I must find a way into that Super Computer complex. Maybe there's something on the dispatch control panel that will help me get in. Gee, what's that big red thing overhead? It looks pretty much like nothing to me, so why the heck did they put it there? Just another typical, pointless Vohaul thing.

There's nothing on this computer screen either. Darn it, how do they do it? Well, obviously there's nothing here I can use. I might as well go to the east room and try to figure out something there. Hopefully that Sequel Police guard will be gone.

At the Super Computer Entrance

Great! No Sequel Police in here this time. Now, let's see if I can get this door open. I'm positive this is a tunnel entrance leading to *something*. Dollars to doughnuts this is the entrance to the Super Computer complex itself. Darn it! This door won't budge! Now what am I going to do? I have to get inside that complex. I just know the key to stopping Vohaul is somewhere in there—it has to be!

Well, I can't do anything else here until I figure out how to get through that door. I need something that will cut through steel. Gee, I'm just drawing blanks. What in the world would be strong enough to cut through a steel lock? Let's see, I don't have a blow torch, and I don't make a habit of carrying around acid in any form . . . until *now*, that is. That acidic green blob that was after me in the sewer tunnels—I'll bet that stuff is strong enough to cut through anything, except glass, of course. OK, so maybe it isn't all that strong. It could have cut through *me*, though, that's for sure! What the heck, it's worth a try.

OK, green slime, do your stuff. . . . Hey, I think it's actually working! Yes! All right! Open Sesame! Yeah, I did it! I'm in! . . . And I was right, this *is* a tunnel of some sort.

In the Laser Tunnel

Hmmm, there's a keypad here on the wall, whatever that's for. Gee now, what are those rings with the funny holes lining this tunnel? Hmmm, if I didn't know better, I'd think this tunnel was protected by laser beams. . . . Oh my gosh! What if that's what they are—infrared laser beams! Boy, I could have bought the farm right here and now if I had walked through there to the other end! Never, ever rush in where angels fear to tread! That's what my mother taught me, and I've never forgotten it—well, except on occasion.

So, isn't this a fine kettle of fish. I'll bet that keypad on the wall sets the angles for the laser beams—only I have no way of telling where they are and where they should be. I need some infrared goggles or something. Surely there's a pair lying around here somewhere. . . . Oh right, Wilco—they're three for a buckazoid in the gift shop next door!

Boy, this is a real dilemma. I've come across some puzzling circumstances before, but nothing compared to this. My wit and cunning have always seen me through, and now I'm stumped.

Darn it, just a few moments ago I was all hyped up about saving Xenon. Now I haven't the faintest idea what to do. This is just too taxing. My poor brain needs a rest. Maybe if I clear my mind of everything bad that's happened and try to think more pleasant thoughts, something will come up. But what should I think about? Gladys and Fly-Boy? No, that just makes me sad. What have I experienced lately that really gives me a lift?

Zondra! That's it! When I saved Zondra and all her Latex friends from that giant sea slug. . . . Yeah. . . . Boy, that made me feel like a real he-man. I was on top of the world the moment those gals came racing over to thank me for saving their lives. What a great feeling! I put a definite end to that slugger, and so cleverly. . . . Yes, sir, give that man a cigar!

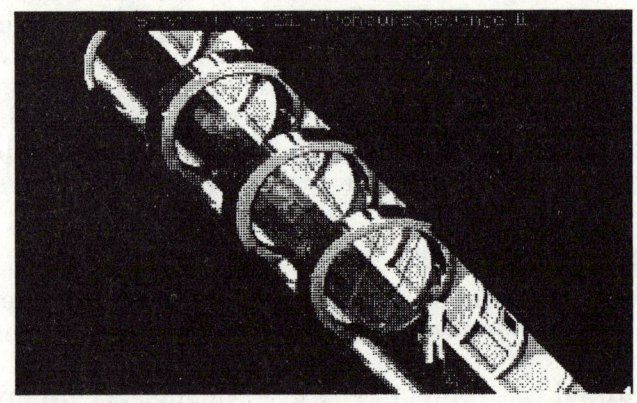

Hope you remember your eighth-grade geometry. You're going to need it.

Hmmm, speaking of cigars, I do believe I deserve one—even if it did fall out of that snoutface's mouth. I'll just wipe it off real carefully. . . . Darn it, I never did go back and put it in one of his burgers. Heh heh heh, wouldn't that have been a hoot?! Yeah, I'm feeling much more relaxed already. I'll just light this cigar, take a few quick puffs to get it going, and . . . (*cough cough cough*). . . . Ugh! Gag! I should have known that oaf wouldn't know a fine cigar from tumbleweed . . . (*cough cough*). . . . They actually call this a cigar? That's the worst thing I've ever tasted! Yuck!

Holy smoke, I don't believe it! I've been witty and cunning once more and I didn't even know it! I didn't even try! I can actually see the laser beams now! This is fantastic! Now I can use the keypad on the wall to adjust their angles, and I'll be back in business in no time!

Inside the Complex

Wow! I still can't get over what just happened! What a coup! . . . Oh man, would you get a load of this place. This is unbelievable. I knew this structure was big, but I never dreamed it would be anything like this! Oh my gosh, what are those disgusting growths all over the hardware? It must be some grotesque part of Vohaul and his virus. He really is the scum of the universe.

Uh oh, what's that noise? It sounds like . . . humming! Y-y-y-yikes! I'll bet it's another one of those death droids! I better get the heck back into that tunnel—and fast!

Wide is the gate and broad is the way that could lead to your destruction. Remember to remember!

Whew! That was close. I hope he's gone now, but there's only one way to find out. . . . All's clear so far, but I have a feeling that won't last for long. There must be some kind of movement detector that alerts the droids when someone enters the complex. Now how am I ever going to get anywhere with those things floating around?

Say, what's that terminal-looking thing over there? Hey, there's one on every level. They're all over the place, as a matter of fact. They're obviously significant, but for what? I need to take a closer look. I'll be swift about it for sure. . . . Hmmm, now *that's* interesting. It's obviously some sort of plug. . . . Uh oh, I hear that humming again. I'm outta here!

Back in the Laser Tunnel

Boy, another close call. I wonder if my PocketPal would work if I found a way to plug it into that terminal. Heck, it's worth a try, and it would probably give me a lot of information. Now all I have to do is figure out *how* to plug it in. Let me think. . . . Wait a minute! I saw some PocketPal adapters in the RadioShock automated catalog! Well, I didn't actually *look* at the adapters—not after seeing their prices—but gosh, it's probably worth it for me to go back to check into buying one. I know I have enough buckazoids left. Surely they have a configuration that will work the plugs on those terminals.

Why am I wasting my time here thinking about it? I have to get back to the time pod on the double!

In the Time Pod

Thank goodness no Sequel Police were hanging around this time. I've been lucky about that so far, but I may not be so lucky where I'm going. Let's see, I know I wrote down those mall coordinates—and boy am I glad I did. I just thought they might come in handy sometime.

Here they are. . . . I just hope I don't run into any Sequel Police at the mall!

Game Points Earned

Action	Points
Correctly rotating the laser beams	15
Lighting the cigar	10
Opening the tunnel door	10

Return to Space Quest X

Mission: Purchase the correct PocketPal adapter.
Total Points: 15

Galaxy Galleria Mall

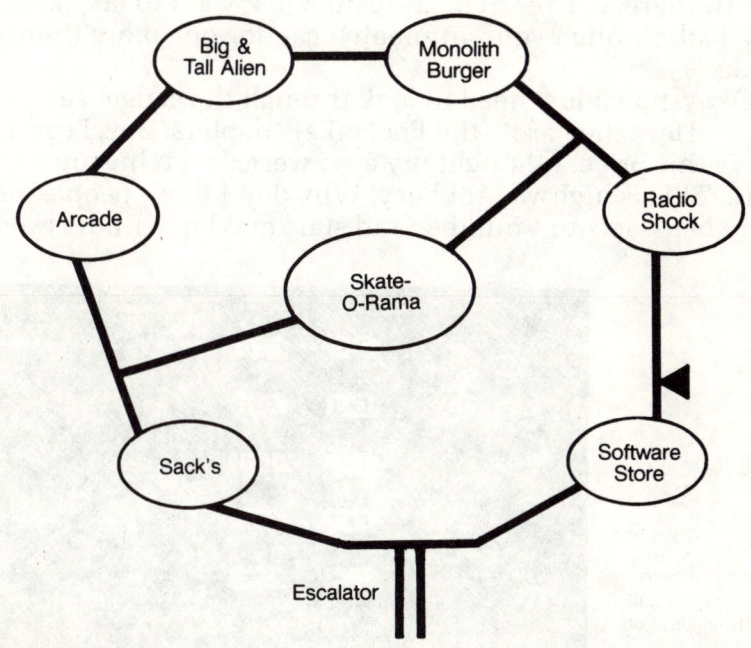

Inventory

PocketPal adapter

In the Arcade

I don't see any Sequel Police right now, but dollars to doughnuts they're lurking around here somewhere. I'll just make
this a very short trip—no stopping to window shop or anything.

Gee, I just hope those guys aren't waiting at the time pod when I get back. I'd hate to have to go through the Skate-O-Rama again.

RadioShock (The Dandy Company)

Well, I'm glad the mall stays open late—not that I have any idea what time it is. It just seems to have been a very long day. "Hi there. . . . Yes, it's me again. I just want to take a quick look through your automated catalog one more time. Thanks."

Okay, now I just need to look through the gadget section. . . . Here they are—the PocketPal adapters. Boy, I can't believe this price. I thought my eyes were deceiving me before. This is highway robbery! Why don't these people just get my bank account number and start making withdrawals!

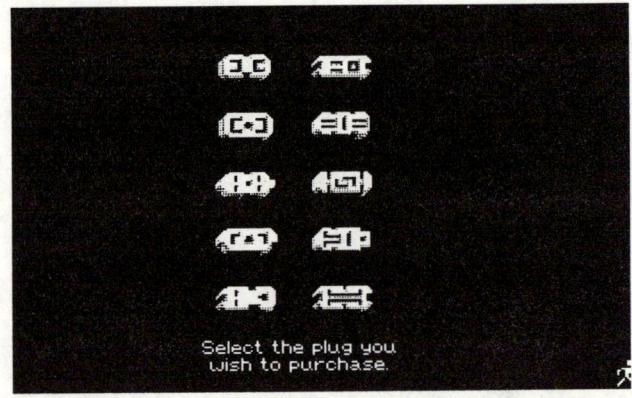

Eeny meeny miny-. . . . Sorry, Roger. That just won't get it this time. Did you remember to remember?

Wow, look at all the adapters. Who would have thought there were so many different ones for a single little Pocket-Pal? I'm just glad I have a good memory for configurations. Guess that comes from my years of being a sanitation engineer and learning all the sewer duct routes. There's the one! That's the adapter I need! Oh, I'm so thankful they actually have it in stock!

Boy, it really pains me to pay for this—I'm going to be wiped out clean, but I'm desperate. Ha—what was that I said earlier about having to be desperate to pay these prices? Well I'm desperate now, so here goes. . . .

"You people ought to be ashamed of yourselves for pricing merchandise so high—this is price rigging if I've ever seen it! Did you know this one little adapter may well be the key to saving the universe? . . . What do you mean it ought to be worth it then? It's a piece of rubber, you clown! Next time, I'll take my business to Circuitry City!" Dumb robots. What do they know about anything anyway? They just do what they're wired to do.

Now I have to get back to that time pod before the Sequel Police discover it!

In the Time Pod

Whew! I made it once more! I just keep wondering when my luck is going to run out. No—I can't think that way. I have to think positively! Now that I have this adapter, I should be able to plug into that computer terminal and at least figure out something. I don't know what I'll figure out, but it will be something.

Where are those darn Xenon coordinates now? I know I stuck them in my extra pocket—it's hard to keep these things straight. Here they are. . . . No, those are the ones to Ulence Flats. Yikes! I *definitely* don't want to go there again! . . . Here they are. Now I can be on my way. Good-bye, mall. If I'm successful, I won't be seeing you for a long, long time!

Game Points Earned

Action	Points
Buying the correct adapter	15

Object Locations

Object	Location
PocketPal adapter	RadioShock

Space Quest XII: The Sequel

Mission: Wipe out Vohaul from the Super Computer system—and save Junior.
Total Points: 123

Xenon Super Computer Complex

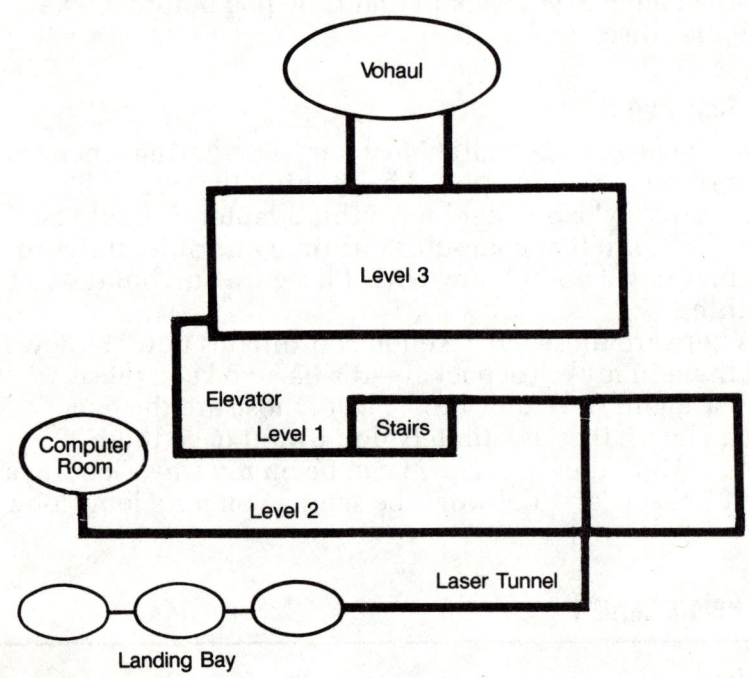

Inventory

floppy disk

At the Time Pod Dispatch Area

No Sequel Police again, thank goodness. This is almost too good to be true. Now if my luck will just hold out long enough for me to get to one of those terminals. . . . I have no idea what's going to come up on my PocketPal screen, but hopefully it will be information I can use.

Super Computer Entrance Area

Well, I was able to make it all the way through the time pod dispatch area, the landing pod bay, and the laser beam entrance tunnel without incident. I can't believe for one minute, though, that Vohaul doesn't know I'm here.

He must be waiting for something—everything's falling too neatly into place now. I wonder when he'll pull the big punch. It has to be soon—I'm getting too close. If he doesn't know *I'm* here by now, he will very soon. This time I'm pulling out all the stops. This is going to be no-holes-barred vengeance and retribution! Vohaul will pay for what he did—or rather, for what he *thinks* he did—or rather, for what he thinks he's *going* to do. I guess it depends on which time period you're in. Anyway, the point is—he's going to regret everything!

At the First Terminal

Now to plug in my handy-dandy PocketPal to this terminal. Hmmm, what does that fine print say there at the bottom? *U-Plug-It Terminal, Ronko Corp.* Oh brother, that figures. . . . I better be swift about this before one of those humming death droids decides to bring his song-and-dance routine over here.

Ack! Look at that! Those flashing lights—Aliens! Ack! They're closing in on me! Ahhhhh! I've got to get out of here, but which way can I go? There's one moving just to the west of me! He's closing in! He's almost on top of me! Ahhhh! . . . *(puff puff)*. . . . I don't know where I am, but I think I'm running north . . . *(puff puff)*. . . . I'll run down these stairs . . . *(puff puff)*. . . . Oh, please don't let there be one of those things hiding around the corner . . . *(puff puff)*. . . .

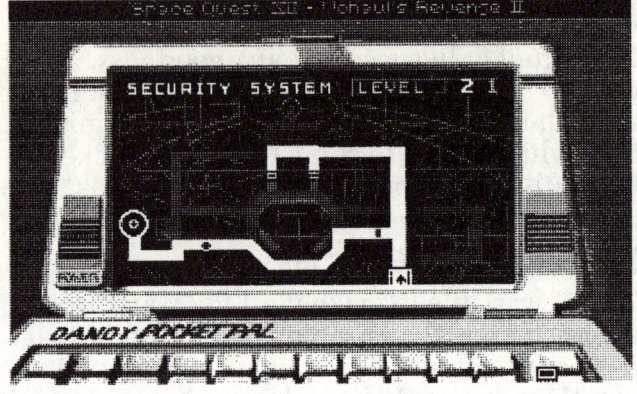

You're not alone, Roger. Better plug in your PocketPal and follow the yellow brick—well, whatever it is—road!

At the Divided Stairway

Whew, I made it . . . *(puff puff)*. . . . Now if I can just catch my breath. I know that . . . that *thing* is still coming after me. Those flashing lights. . . . I know what Aliens look like on a computer screen. That's exactly what they are! Not only do I have to worry about running into one of those humming death droids, now I have Aliens to worry about! Talk about the heebie-jeebies! Boy, Vohaul knew what he was doing this time. He must have seen that movie, too.

Oh no—I see a death droid! He's on the upper level. . . . He doesn't see me . . . thank goodness. I have to get to another terminal fast! I have to know where those Aliens are! I'll just quietly sneak up these stairs and. . . . Huh! Another death droid. . . . Boy those things are loud. Why don't they just scream out an announcement—DEATH DROID COMING, DEATH DROID COMING. It would have about the same effect, for crying out loud.

Good, he's going down the stairway. Now is my chance to get the heck out of here! I need to find another terminal as quickly as possible. At least the coast is clear—no Aliens in sight! Of course, I know that doesn't mean anything—they have a way of closing in when you least expect it.

I could run back the way I came in! Yeah, I could plug in my PocketPal at that first terminal again. . . . It's worth a shot. I *have* to know what's happening!

Back at the First Terminal

OK . . . *(puff puff)*. . . . I wish I could stop long enough to catch my breath . . . *(puff puff)* . . . but there's no time to waste! I'll just plug this in . . . *(puff puff)* . . . and see what's going on! OH NO! They're still after me! They're on to me for sure! I'll bet they can even tell which U-Plug-It terminal I'm using! . . . I must keep going . . . *(puff puff)*. . . . At least nothing was blinking in front of me . . . *(puff puff)*. . . . I'll run as far west as I possibly can. . . . I thought I saw what looked like a room on this end . . . *(puff puff)*. . . . I just wish I had been able to get a better look at the layout of this place on my PocketPal screen.

At the Programming Chamber

I was right . . . *(puff puff)* . . . there is a room here . . . *(puff puff)* . . . but it's behind that door. Oh! Whew! I need to rest— but I can't. I better go out there and find out where the Aliens are. Oh no—a horrible thought just occurred to me! What if they're waiting for me behind that door? After all, they can travel through every kind of pipe and air duct known to man—why wouldn't they be hiding?

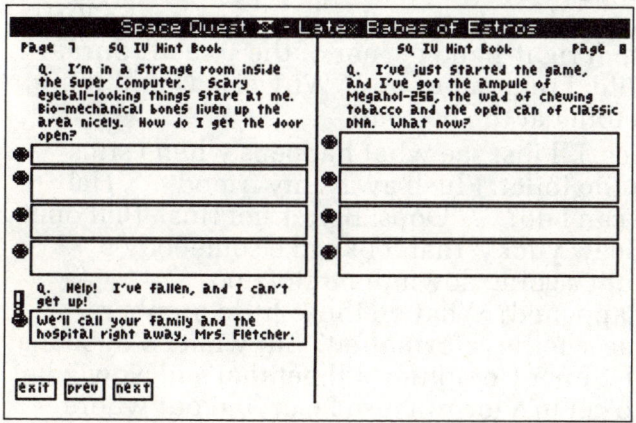

Great hint, Roger.
No one fools you.

Oh gee, I have to get to another terminal! But what if they're out there waiting for me round the corner? Oh gosh, I don't know what to do! What should I do? . . . My hint book! Maybe it has a clue I can use! Oh what did I do with that thing? Here it is. . . . Let's see. . . . Darn it, these pages aren't even in order! I knew this cheap. . . . Hey, wait a minute! Here's a question that sounds like it could apply to my situation! Now, what does the hint say? . . . Arrrgggg! If they don't knock it off with the cutesy answers! Where's the real answer?

Here it is! It says I should punch in this code to open a door. What the heck—I might as well try. The only thing I'll lose if I don't is my life! Well, here goes . . . and those last two digits are 6 and 9. Now press Enter. . . . It worked! Yes! It worked!

Inside the Programming Chamber

Now what do I do? Hmmm, I recognize this setup. It's a menu for some type of computer program. Whoa, I wonder if this is THE computer program—the one that runs everything. What do those little pictures mean? Well I know what *that* one means—obviously someone was goofing off. Why else would there be a picture for a *King's Quest* game?

Gee, that one looks like a little droid. Hmmm, it almost looks like one of those death droids. Hey, wait a minute! I know what these are—they're called icons! Gladys used to have stuff like this on her computer screen all the time! Now what was it I saw her do to get rid of the things she didn't want? . . . Didn't she drag the pic—I mean icon—to the little thing in the bottom right-hand corner of the screen? Sure, that's what she did. I remember now! Why else would there be a picture of a toilet at the bottom?

OK, here goes. I'll just see what happens when I stick that droid down the toilet. Flush away, my friend. . . . Ha! Now which one can I do? . . . Oops! Better not flush *that* one! What else? . . . Eeew yucky, that looks like somebody's brains. I'm afraid it's going down the toilet, too.

Hey, what happened? What do those little numbers mean? Something's being reformatted? But what? Maybe it's the entire Xenon Super Computer! I'll bet that's it! Now what do I do? I need to get to a terminal so I can find out where those Aliens are! I just hope I got rid of the death droids when I flushed their icon down the toilet.

Outside the Programming Chamber

Whew! I don't hear any death droids humming about—but those Aliens! They could be closing in on me right now! Oh, I don't want to die like that! I have to know what's going on!

There's a terminal. . . . I'll plug in my PocketPal as quickly as possible, and. . . . "Ack! Vohaul! It's you, you low-down, good-for-nothing sewer scum! I'm going to stop you, Fatback, if it's the last thing I do! You can gloat all you want, but I can promise you this—I'll huff, and I'll puff, and I'll blow your entire insides out if that's what it ta. . . . Huh? What are you doing with that guy? I recognize him! He's a Time Ripper. He was part of the Xenon resistance! He helped me escape from those buffoons you call police. . . . He's WHAT? . . . *(gasp)* . . . My *son?!*

Garbage in, gar-bage out. But he does have some pretty good scoop to tell.

"Vohaul! Wait a minute! You come back here you gutter slime! I'm not through with you yet!" My son—Roger, Jr.! I have a son, and I didn't even know it because this is really the future and it hasn't even happened yet! Gosh, I wonder if Gladys is his mother. . . . I have to save him! That's all that matters now! I have to save my son! Vohaul, you're as good as dead! If he's intricately tied to this Super Computer the way I think he is, I'll bet that reformatting sequence is going to reformat him! Oh I do hope so.

Oh no! In my utter shock and surprise over seeing Vo-haul's disgusting face and learning about the son I never knew I had, I completely forgot about those darned Aliens! This is really strange, though. I *know* that's what those flash-ing lights on my screen signified. What else could they have been? I'll go to another terminal and find out; I don't want Vohaul's ugly head popping back up again.

Terminal at the Steps

There's another terminal—the first one I used. I think I'll by-pass that one just in case I'm being followed. I remember seeing one at the steps leading to the first level, though. I'll try my PocketPal there. I just hope those Aliens don't catch up with me.

Here we are—now let's see what happens this time. . . . Hmmm, no flashing lights. That's really odd. . . . I just knew there were aliens after me. Hey, wait a minute—now I get it! Those flashing lights weren't Aliens after all—they were the death droids! Of course! And now they're all gone because I flushed their little main icon down the tank! Very good—now I can get back to the business at hand without constantly looking over my shoulder.

Saaay, I didn't really have time to notice this layout of the Super Computer before. I was too busy running from death droids and phantom Aliens. Now that I have time to study it, I can see there are three distinct levels here. It looks like I'm headed in the right direction, too. That area on the third level—I'll bet that's where Vohaul has my son—and apparently it can only be reached by elevator. If my eyes weren't deceiving me, there's a glass tube elevator on the first floor—exactly where I'm headed! Hang on, Junior! Dad's on his way!

Outside the Elevator, Level 3

I was right; this elevator is the only way to level 3. That little PocketPal has really come in handy. I guess the adapter plug was worth all those buckazoids after all; otherwise, I wouldn't even know my own son existed.

What was that? Some voice just announced something about that reformatting sequence again. Whatever I set in motion in that programming room is apparently still going on. I just wonder how much time Roger, Jr. has left. I have to get to him before it's too late!

I'm sure this is the right direction. What else would be on the other end of such a huge tunnel but the titanic Vo-haul? All I know is he better not have harmed my son!

The Hologram Pedestal

"Junior! It's you! I knew I'd find you. . . . Huh? What do you mean you're Vohaul/Roger, Jr. and I'm not? I know I'm not, but why are you talking like that? Son, it's me, Roger, your dad! Can you hear me? Why did you throw that disk away just now? What was on it?

"I don't know what's going on here—I'm totally confused. Junior, what's wrong with you? Why do you want to fight with me? . . . Listen to me. . . . Ouch! . . . I'm your dad. . . . Ooo! . . . Why won't you reason with me? . . . Uhhh! . . . Something's . . . Ooof! . . . wrong! This can't be Junior! It's got to be Vohaul. He's inside my son's body! Why you. . . . Ooo! . . . I'll kill you, Vohaul! Take that! . . . Now what have you done with my son? Where is he? You may have his body, but you'll never have his soul!"

The computer platform! He must have stepped on the button that raises the console! What does all this mean? Beam upload, beam download? . . . What does this mean, and what does it have to do with a reformatting sequence?

That disk! I have to get that disk. I don't know how much time I have left, but I think it's becoming critical! I should be able to reach it from that ladder. Oh, just hang in there a little while longer, Junior. Dad will have you back to your old self in no time!

I've got it! I have the disk, and with no time to spare. . . . Now I have to insert it in the disk drive. . . . That's it! That's how he did it! Vohaul downloaded his essence into my son's body, and Roger, Jr.'s essence must be on this floppy disk! How DARE he throw away my son's essence!

Now I have to upload that fool back into the Super Computer before it completely reformats. Where's the button? There it is! . . . It's working! . . . He's uploading back into the Super Computer, and when the reformatting sequence is complete, he'll be erased like data on a giant disk! "You thought you were the future, Vohaul, but the truth is, you're HISTORY!"

My son! Now I have to restore Roger, Jr. to his body! The only way I can do that is to upload him from this disk—but do I have time? I have no choice—there's no other way. I have to do this before the reformatting is complete. He's uploaded. . . . I'm running out of time. . . . Oh hurry, Wilco, for once in your life, don't goof it up. . . . And now to download him from the Super Computer's beam. This has to work! It has to! Come on . . . work!

"Son! Is that you? . . . It *is* you! You're alive! Oh, Junior! We did it—I mean *I* did it! You're saved, thanks to me!"

After the Rescue: A Family Reunion

Sorry, but just as in *The Pirates of Pestulon*, this part's a secret! You'll have to rescue Junior and destroy Sludge Vohaul if you want to witness the touching reunion and subsequent farewell of father and son. Good luck! See you in *Space Quest ??*

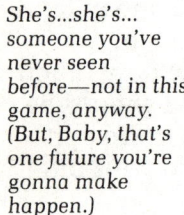

She's...she's... someone you've never seen before—not in this game, anyway. (But, Baby, that's one future you're gonna make happen.)

Game Points Earned

Action	Points
Activating the formatting sequence	15
Deactivating the death droids	5
Downloading Junior to his body	10
Entering Vohaul's chamber for the showdown	5
Finding the formatting chamber	10
Formatting over Vohaul	25
Getting Junior's body into the beam	15
Hooking up the PocketPal	10
Inserting the disk in the computer	5
Installing the battery in the PocketPal	3
Opening the formatting chamber door	10
Plugging the adpater into the PocketPal	3
Retrieving the disk	5
Taking the battery out of the rabbit	3

Object Locations

Object	Location
Floppy disk	Vohaul's pedestal

Appendix

You should be ashamed of yourself, giving up so easily. This wasn't that difficult!

PLEH SU! EW ERA GNIEB DLEH EVITPAC YB TFOSMUCS NO EHT LLAMS NOOM FO NOLUTSEP. NA ELBARTENEPMI ECROF DLEIF SDNUORRUS EHT NOOM. TI TSUM TSRIF EB DETAVITCAED. STI NIGIRO SI NWONKNU OT SU. TFOSMUCS YTIRUCES SI DEMRA HTIW OLLEJ SLOTSIP. ER'EW GNITNUOC NO UOY REVEOHW UOY ERA.

OWT SYUG NI ELBUORT

Index